娃娃服飾縫紉書

HANON

—————— 應用設計篇 ——————

藤井里美

contents

arrangement

本書是娃娃尺寸的洋服縫製技法書【HANON：娃娃服飾縫紉書】系列第二本作品。

相較於第一本作品，內容更加單純、簡單，以初學者也能夠容易縫製的版型為基礎，讓大家

可以依照自己的喜好，將衣身、衣領、衣袖、下襬等部位自行調整設計。

S尺寸的ruruko、JERRY BERRY、Popping等等；

M尺寸的Neo Blythe及b.m.b.Cherry等等；

L尺寸的U-noa Quluts Light及momoko等等；

本書收錄了這些1/6尺寸的3種不同類型的紙型。

說明雖然是以使用縫紉機的機縫為主，但也可以使用手縫製作。

縫製的方法有很多，請各位試著以適合自己的方法製作即可。

希望本書能讓大家有所參考。

為了要讓大家都能享受到服裝搭配的樂趣。

書中特別準備了數種不同長度的裙子及圍裙。

同時也有尺寸方便重疊穿搭的大衣、夾克外套等等。

各位也可以和我第一本作品中刊載的衣服、襪子、鞋子、布娃娃等配件，一起與服裝搭配。

請把HANON的風格收進各位的娃娃衣櫃之中吧！

如果我能夠協助各位對自己的愛娃表達更深切的愛意，那便是我的榮幸了。

This book is the second edition of DOLL SEWING BOOK HANON,
which is for making doll-sized clothes. Based on a simpler and easier pattern
than the first one, you can enjoy your bodice, collar, sleeve, skirt, etc.
by combining your favorite arrangements.

S size is for ruruko, JERRY BERRY, Popping
M size is for Neo Blythe, b.m.b.Cherry
L size is for U-noa Quluts Light, momoko.

The book describes using sewing machine,
but I think you can also sew by hand-sewing reverse stitching.
There are various sewing methods, but please try to make clothes
by a sewing method that suits yourself. I hope this book helps you.

In order to let you know the enjoying of coordination,
I have prepared several length of skirts and aprons pattern.
Please enjoy coordination with clothes, socks, shoes, and stuffed toys
that are published in the first edition.

Please add the HANON style to your doll's wardrobe,
I would be glad if I could help you more shower your dolls with love.

HANON
Satomi Fujii

M size Basic Darts Dress (Arrangement), Apron Dress (Arrangement) & Hat

M size Basic Darts Dress (Arrangement) & Bag

M size Basic Darts Dress & Detachable Collar

M size Basic A-line Dress (Arrangement) / M size A-line Dress (Arrangement), Hat & Bag

M size Jacket, Basic Blouse (Arrangement), Knickerbockers, Hat & Bag

M size Basic A-line Dress (Arrangement)

M size Basic Blouse (Arrangement), Detachable Collar, Knickerbockers & Hat / M size Coat & Hat

M size Basic A-line Dress (Arrangement), Apron Dress & Bag

M size Basic Blouse (Arrangement) & Knickerbockers / M size Basic Blouse (Arrangement), Tiered Skirt & Apron Dress (Arrangement)

M size Basic A-line Dress (Arrangement) & Hat

S size Basic A-line Dress (Arrangement)

（上）*S size Basic Darts Dress (Arrangement)* /（下）*S size Coat & Basic Darts Dress (Arrangement)*

S size Jacket & Tiered Skirt

S size Basic Blouse (Arrangement) & Tiered Skirt

S size Basic A-line Dress (Arrangement) & Apron Dress / L size Basic Darts Dress (Arrangement) & Apron Dress

S size Basic Blouse (Arrangement), Knickerbockers, Hat & Bag

S size Basic Darts Dress (Arrangement)

S size Basic Darts Dress (Arrangement) & Hat
L size Basic Blouse (Arrangement), Tiered Skirt, Hat

L size Basic A-line Dress (Arrangement) & Detachable Collar (Arrangement) / L size Basic A-line Dress (Arrangement)

L size Basic Darts Dress (Arrangement) & Apron Dress

L size Jacket, Knickerbockers & Hat / L size Coat, Basic Blouse (Arrangement), Knickerbockers & Hat

Tools

在開始縫製娃娃衣之前，先把工具收集齊全吧！
有一些平常洋裁不怎麼會使用到的工具，對於製作小尺寸娃娃衣來說，
卻能發揮相當便利的作用哦，非常推薦給大家使用。

絲質緞帶 *Embroidery Silk Ribbon*
緞帶刺繡用的3.5mm寬緞帶，質地柔
韌方便使用，而且產品種類也很豐富。

刺繡線 *Cotton Embroidery Floss*
本書使用的是DMC的25號線。

拆線刀 *Seam Ripper*
如果縫線縫歪了的話，要果斷地用拆
線刀把縫線拆掉重新縫過。

鉗子 *Pliers*
本書使用的是手藝用的小鉗子，要將
小塊布片翻回正面時非常好用的工具。

修線剪刀 *Thread Scissors*
用來修剪手縫線與機縫線的線頭及尾
端。

頂針 *Thimble*
在衣服加上刺繡裝飾，或是線縫的時
候使用。

錐子 *Tailor's awl*
可以用來將布料翻面，或是將縫合的
邊角呈現出來。使用縫紉機縫製時，
也可以用來按壓固定布料。

裁縫剪刀 *Dressmaking Scissors*
請選用鋒利好裁剪、適合細微作業的
小尺寸剪刀。本書使用的是美鈴的褙
棉布剪刀。

縫線 *Sewing Thread*
不管是機縫或是手縫，我愛用的品牌
"都是TicTic PREMIER"。

布用接著劑 *Fabric Glue*
如果是暫時固定用的話，本書使用的
是KAWAGUCHI的布用接著劑。如果
需要緊密固定的話，建議使用皮革用
的接著劑。

防綻液 *Fray Stopper*
KAWAGUCHI的Pique防綻液是我的

愛用品。布料裁切後，要在端面塗抹
上防綻液處理避免綻開。

粉土筆 *Tailor's Chalk*
較薄的布料使用不易滲到背面的
KARISMA FABRIC布用自動鉛筆；
較厚的布料使用Cosmo的Chacopaper
極細水性粉土筆；深色布料使用
CLOVER的白色熱消粉土筆。

蕾絲 *Laces*
本書的範例使用的是復古蕾絲。新的
蕾絲布如果顏色太鮮艷漂亮，不容易
和其他布料搭配的話，可以先用草木
染或是紅茶染處理成自己喜歡的顏色
後再搭配到衣服上。

暗鈕 *Snaps*
本書使用的是5mm的圓型暗鈕。

縫針　珠針　絲針　定規尺
Handsewing Needles, Dressmaker Pins, Silk Pin, Ruler

Basic A-line Dress

基本的A字連身裙

這是只要3張紙型就能夠製作，非常推薦給裁縫初學者的基本型連身裙。

然後再依照不同技術等級，一步一步追加衣領、衣袖、下襬、衣身的應用設計吧！

平織薄棉布 〔衣身〕	S 20×32cm M 20×38cm L 25×40cm	平織薄棉布 〔衣袖〕	S 12×22cm M 12×24cm L 15×28cm	裡布 〔尼龍紗〕	S,M,L 7×7cm
平織薄棉布 〔衣領〕	S 4×20cm M 4×20cm L 4×23cm	平織薄棉布 〔下襬荷葉邊布料〕	S 4×60cm M 4×60cm L 4×68cm	暗釦	S、M、L 2 組

1

依照紙型將各部位的布片裁切下來，布端塗抹防綻液進行處理。
→「衣身」的應用設計請參考第79-83頁。

2

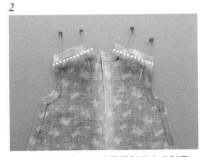

將前衣身與後衣身以正面相對的方式對齊，縫合肩部。

3

將縫份以熨斗燙開。
→如要加裝「衣領」，請參考第84-88頁。

4

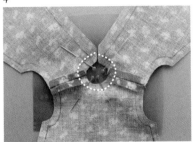

準備一塊粗裁成7×7cm左右的裡布，與衣身以正面相對的方式重疊後，將頸部周圍縫合。

5

將裡布如照片般裁切，在頸部周圍的縫份剪出細密的牙口。請小心不要剪到縫線。

6

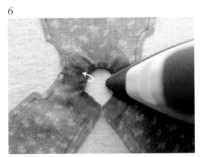

將裡布翻過來，用熨斗燙平。

7

將後開口的縫份，斜折至「開口止點」記號的稍微下方。

8

沿著「開口止點」的稍微下方～頸部周圍「開口止點」的稍微下方車縫一道壓線。
→如要加裝「衣袖」，請參考第89-95頁。

9

在袖襱的縫份剪出細密的牙口。

Let's enjoy that collars, sleeves and hem arrangements
according to your level.

1. Arrange the paper templates on the fabric and cut all the sections, then apply fray stopper liquid to all the edges.
[refer to p.79-83 for the front arrangements]   *2.* Match the right sides of the front and back by the shoulders and sew.
3. Iron open the seam allowances. [refer to p.84-88 for the collar arrangements]   *4.* Cut the dough about 7cm square for lining.
Match the bodice and the lining and sew the neckline.   *5.* Cut the lining as shown and cut slit in the seam allowance of the neckline.
Be careful not to cut the stitches.   *6.* Turn the lining right side out and iron.
7. Fold the back opening inwards slightly below the opening stop marker and iron flat.   *8.* Sew the edges from the stop marker along the neckline,
then to the stop marker. [refer to p.89-95 for the sleeve arrangements]   *9.* Cut fine slits into the seam allowance of the armholes.

10

用熨斗將縫份燙平,再以布用接著劑暫時固定。

11

將袖襱縫合起來。

12

將前衣身與後衣身以正面相對的方式對齊,縫合兩側脇邊。

13

用熨斗將縫份由中間燙開。

14

將下襱的縫份折向內側,用熨斗燙平固定。
→「下襱」的應用設計請參考第96-99頁。

15

將下襱縫合起來。

16

將後開口以正面相對的方式重疊,從「開口止點」開始縫合到下襱。

17

翻回正面,用熨斗將縫份由中間燙開。

18

裝上暗釦,這樣就完成了。

10. Fold the seam allowances and apply fabric glue to the seam allowance.   11. Sew the armholes.
12. Match the right sides of the front and back bodice sections facing, and sew them together.
13. Iron open the seam allowance.   14. Fold the seam allowance of the skirt hem inwards with an iron. [refer to p.96-99 for the hem arrangements]
15. Sew the hem.   16. With right sides facing, sew from the stop marker to the hem.   17. Iron open the seam allowance.   18. Add snaps to the back opening.

Basic Blouse

基本的罩衫

製作方法幾乎和A字連身裙一樣。下襬是否有碎褶設計也會帶來不同的印象。

如果沒有碎褶，只要再加上荷葉邊就可以變身為Peplum腰間裝飾罩衫。

請自由發揮各種不同的應用設計吧。

平織薄棉布	S	10×24cm	平織薄棉布	S	12×22cm	裡布	S,M,L 15×15cm
〔衣身〕	M	10×26cm	〔衣袖〕	M	12×24cm	〔尼龍紗〕	
	L	11×28cm		L	15×28cm		
						暗釦	S,M,L 2組
平織薄棉布	S	4×20cm	平織薄棉布	S	4×36cm		
〔衣領〕	M	4×20cm	〔下襬荷葉邊	M	4×38cm		
	L	4×23cm	布料〕	L	4×40cm		

1

依照紙型將各部位的布片裁切下來，布端塗
抹防綻液進行處理。
→「衣身」的應用設計請參考第79-83頁。

2

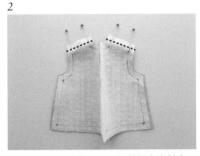

將前衣身與後衣身以正面相對的方式對齊，
縫合肩部。

3

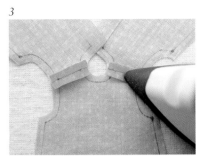

將縫份以熨斗燙開。
→如要加裝「衣領」，請參考第84-88頁。

4

準備一塊粗裁成15×15cm左右的裡布，與
衣身以正面相對的方式重疊後，以珠針暫時
固定。

5

沿著後衣身的下襬～頸部周圍～後衣身的下
襬縫合起來。

6

將裡布如照片般裁切，在頸部周圍的縫份剪
出細密的牙口，將邊角修掉。請小心不要剪
到縫線。

7

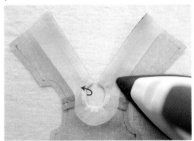

將裡布翻過來，用熨斗燙平。
→如要加裝「衣袖」，請參考第89-95頁。

8

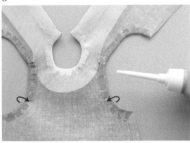

在袖襱的縫份剪出細密的牙口，用熨斗將縫
份燙平，再以布用接著劑暫時固定。

9

以縫紉機將後衣身的下襬～頸部周圍～後衣
身的下襬，以及袖襱縫合起來。

It is very easy to make it as Basic A-line dress.
Please enjoy various arrangements.

1. Arrange the paper templates on the fabric and cut all the sections, then apply fray stopper liquid to all the edges.
[refer to p.79-83 for the front arrangements] *2.* Match the right sides of the front and back by the shoulders and sew.
3. Iron open the seam allowances. [refer to p84-88 for the collar arrangements] *4.* Cut the dough about 15cm square for lining.
Match the bodice and the lining. *5.* Sew as pictured. *6.* Cut the lining as shown and cut slit in the seam allowance of the neckline.
Be careful not to cut the stitches. *7.* Turn the lining right side out and iron. [refer to p.89-95 for the sleeve arrangements]
8. Cut fine slits into the seam allowance of the armholes. Fold the seam allowances and apply fabric glue. *9.* Sew the edges as pictured.

10

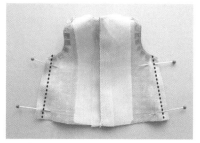

將前衣身與後衣身以正面相對的方式對齊，
縫合兩側脇邊。

11

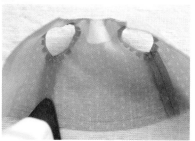

用熨斗將縫份由中間燙開。
→「下襬」的應用設計請參考第96-99頁。

12

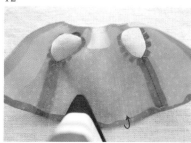

將下襬的縫份折向內側，用熨斗燙平固定。

13

在下襬縫上一道針腳寬度約2.5mm的抽褶
用縫線。
→關於碎褶，請參考第100頁。

14

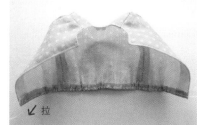

拉動抽褶用的縫線，抽出〔S:11.5cm M:12cm
L:13.5cm〕的碎褶。

15

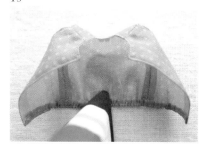

將碎褶間隔整理平均後，用熨斗整燙固定。

16

由正面車縫一道壓線。

17

將抽褶線拔除。

18

在後開口裝上暗釦，這樣就完成了。

10. Match the right sides of the front and back bodice sections facing, and sew them together.   *11.* Iron open the seam allowance.
[refer to p.96-99 for the hem arrangements]   *12.* Fold the seam allowance of the hem inwards with an iron. If you don't want to gather the hem, sew the edges.
13. If you want to gather the hem, Use a machine to sew gathering stitches in the seam allowance of the hem. [refer to p.100 for gathering]
Make the stitch length 2.5mm and sew one lines on the seam allowance.   *14.* Gather the hem to [S:11.5cm M:12cm L:13.5cm] with the finished line.
15. Shape the gathering and iron flat.   *16.* Sew the hem.   *17.* Remove the gather thread.   *18.* Add snaps to complete the blouse.

Basic Darts Dress

基本的打褶連身裙

這是用途廣泛，便於應用設計，在腰間有縫接線的連身裙。

因為裙片使用的是長方形的版型，所以下襬的應用設計相當輕鬆簡單。

平織薄棉布 〔連身裙〕	S 15×50cm M 15×52cm L 17×60cm	平織薄棉布 〔衣袖〕	S 12×22cm M 12×24cm L 15×28cm
平織薄棉布 〔衣領〕	S 4×20cm M 4×20cm L 4×23cm	裡布 〔尼龍紗〕	S,M,L 15×15cm
		暗鈕	S,M,L 2組

1

依照紙型將各部位的布片裁切下來,布端塗抹防綻液進行處理。
→「衣身」的應用設計請參考第79-83頁。

2

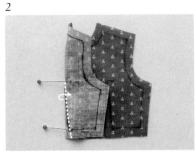

將前衣身的縫合線以正面相對的方式對齊之後,縫合起來。

3

用熨斗將縫份倒向內側。

4

將前衣身與後衣身以正面相對的方式對齊,縫合肩部。

5

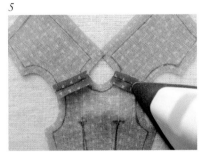

將縫份以熨斗燙開。
→如要加裝「衣領」,請參考第84-88頁。

6

準備一塊粗裁成15×15cm左右的裡布,與衣身以正面相對的方式重疊後,沿著後開口的下襬~頸部周圍~後開口的下襬縫合起來。

7

將裡布如照片般裁切,在頸部周圍的縫份剪出細密的牙口。請小心不要剪到縫線。

8

將裡布翻過來,用熨斗燙平。

9

以縫紉機將後開口的下襬~頸部周圍~後開口的下襬,車縫一道壓線。
→如要加裝「衣袖」,請參考第89-95頁。

The skirt has a rectangular pattern,
so you can easily enjoy the arrangement.

1. Arrange the paper templates on the fabric and cut all the sections, then apply fray stopper liquid to all the edges.
[refer to p.79-83 for the front arrangements] *2.* Fold and sew the darts. *3.* Fold the seam allowances of each darts inward and iron.
4. Match the right sides of the front and back by the shoulders and sew. *5.* Iron open the seam allowances. [refer to p.84-88 for the collar arrangements]
6. Cut the dough about 15cm square for lining. Match the bodice and the lining and sew as pictured.
7. Cut the lining as shown and cut slit in the seam allowance of the neckline. Be careful not to cut the stitches.
8. Turn the lining right side out and iron. *9.* Sew the edges as pictured. [refer to p.89-95 for the sleeve arrangements]

10

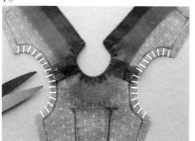

在袖襱的縫份剪出細密的牙口。

11

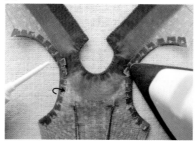

用熨斗將縫份燙平,再以布用接著劑暫時固定。

12

將袖襱縫合起來。

13

將前衣身與後衣身以正面相對的方式對齊,縫合兩側脇邊。

14

用熨斗將縫份由中間燙開。

15

將裙片下襱的縫份折向內側,用熨斗燙平固定。
→「下襱」的應用設計請參考第96-99頁。

16

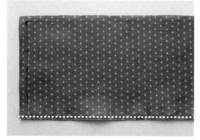

以縫紉機在裙片的下襱,車縫一道壓線。

17

在腰部的縫份縫上2道針腳寬度約2.5mm的抽褶用縫線。
→關於碎褶,請參考第100頁。

18

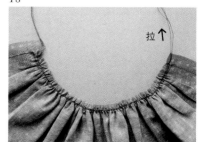

配合衣身腰部的寬度,拉動縫線抽褶,然後再用熨斗整燙固定。

10. Cut fine slits into the seam allowance of the armholes.   *11.* Fold the seam allowances and apply fabric glue to the seam allowance.
12. Sew the armholes.   *13.* Match the right sides of the front and back bodice sections facing, and sew as pictured.   *14.* Iron open the seam allowance.
15. [refer to p.96-99 for the hem arrangements] Fold the seam allowance of the skirt hem inwards with an iron.
16. Sew the hem.   *17.* Use a machine to sew gathering stitches in the upper seam allowance of skirt. Make the stitch length 2.5mm
and sew two lines on the seam allowance. [refer to p.100 for gathering]   *18.* Gather the fabric to match the width of the bodice waist and iron flat.

19

將衣身與裙片以正面相對的方式對齊。此時裙片兩端的縫份會比衣身多出一些。

20

將腰部縫合起來。

21

將縫份倒向衣身那側，用熨斗燙壓固定。

22

將後開口的縫份，由衣身的後開口到「開口止點」的記號稍微下方，斜斜折起。

23

以縫紉機在腰部的衣身那側，車縫一道壓線。

24

這是壓線車縫完成後的狀態。

25

將後開口以正面相對的方式對齊，縫合「開口止點」到下襬。

26

將縫份以熨斗燙開。

27

在後開口裝上暗釦，這樣就完成了。

19. With right sides facing.　*20.* Sew the waist.
21. Fold the seam allowance to the bodice and iron.　*22.* Fold the back opening inwards slightly below the stop marker and iron.
23. Sew the waist.　*24.* Now your stitches are finished.　*25.* With two sides together and sew.
26. Iron open the seam allowance and turn right side out.　*27.* Fasten the snaps at the back opening.

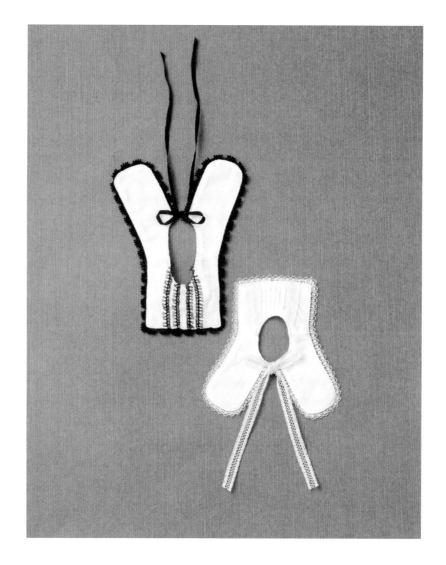

Detachable Collar

裝飾假領

連身裙或是罩衫只要加上裝飾假領來做搭配，整體氣氛馬上就能有極大的改變。
製作所需的布料幅寬較小，耗損的面積也少，很適合用來挑戰有細褶的設計。

平織薄棉布	S 14×20cm	緞帶用蕾絲	S,M,L 15cm×2條
	M 14×20cm	〔5mm幅寬〕	
	L 16×20cm		
		衣身用蕾絲	S 10cm
包邊用蕾絲	S 30cm	〔5mm幅寬〕	M 10cm
〔6mm幅寬〕	M 30cm		L 12cm
	L 35cm		

1

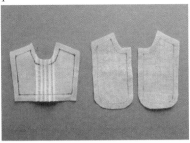

依照紙型將各部位的布片裁切下來，布端塗抹防綻液進行處理。
→細褶的製作方式請參考第82-83頁。

2

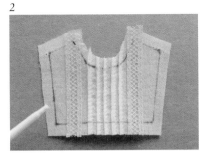

將蕾絲以布用接著劑暫時固定。

3

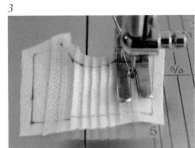

將蕾絲縫合固定。

4

細褶的部分也順便在縫份內車縫壓線。

5

將前衣身與後衣身以正面相對的方式對齊之後，縫合肩部。

6

用熨斗將縫份由中間燙開。

7

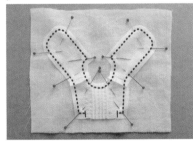

準備一塊粗裁成12×12cm左右的裡布，以正面相對的方式重疊後，保留返口進行縫合。

8

沿著表布形狀，將裡布裁切下來。

9

在頸部周圍及轉角彎弧處的縫份剪出細密的牙口，並將邊角修掉。

Just by matching it to a dress or blouse.
It is recommended for the challenge of pin tucks.

1. Arrange the paper templates on the fabric and cut all the sections, then apply fray stopper liquid to all the edges.
[refer to p.82-83 for the pin tuck arrangement] *2*. Place laces with fabric glue. *3*. Sew the laces. *4*. Sew the seam allowance of the pin tuck.
5. Match the right sides of the front and back facing, and sew the shoulders. *6*. Iron open the seam allowance.
7. Cut the dough about 12cm square for lining. Match the bodice and the lining. Leaving the turn opening and sew as pictured.
8. Cut the lining as pictured. *9*. Cut the corners of the seam allowance, cutting fine slits where the fabric curves.

10

用熨斗將縫份倒向內側。

11

使用鉗子，由返口將布料翻回正面。

12

用熨斗將邊角及彎弧整平。

13

將返口的縫份倒向內側，以布用接著劑暫時固定。

14

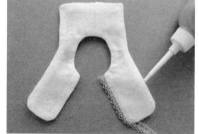

將蕾絲以布用接著劑暫時固定在裡側。

15

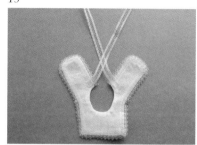

將緞帶用蕾絲〔15cm×2條〕也同樣以布用接著劑暫時固定。

16

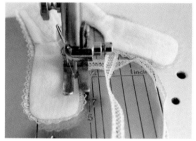

使用縫紉機沿著衣領周圍車縫一道壓線。

17

這麼一來返口就閉合了。

18

用熨斗整平形狀，這樣就完成了。

10. Fold the seam allowance with an iron.　*11.* Turn the right side out using tailor's awl.　*12.* Iron into shape.
13. Fold the seam allowance of the turn opening with fabric glue.　*14.* Place lace on the back side hem with fabric glue.
15. Place the 15cm ribbon on each front opening with fabric glue.　*16.* Sew the edges.　*17.* The turn opening is closed.　*18.* Iron into shape.

Tiered Skirt

百褶裙

這是將2層長方形版型的裙片重疊起來的蛋糕裙款式的裙子。

抽碎褶後用熨斗整平，以及暫定固定處理的這些小小工夫，

可以讓完成後的裙子更加美觀。

平織薄棉布　S　15×40cm　　鬆緊帶　　　S,M,L　15cm
　　　　　　M　17×60cm　〔3mm幅寬〕
　　　　　　L　20×60cm

1

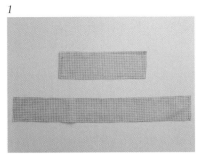

依照紙型將各部位的布片裁切下來，布端塗抹防綻液進行處理。

2

用熨斗將下段裙片(荷葉邊)的下襬縫份折向內側。

3

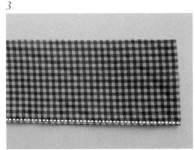

在下襬車縫一道壓線。

4

用熨斗將荷葉邊上部的縫份也折向內側。

5

在腰部的縫份縫上一道針腳寬度約2.5mm的抽褶用縫線。
→關於碎褶，請參考第100頁。

6

配合上段裙片的幅寬，拉出同樣寬度的碎褶。

7

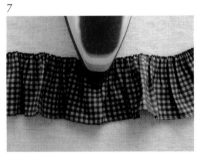

整理碎褶的間隔，用熨斗整燙固定。

8

在上段裙片的下襬正面那側薄薄地塗上一層布用接著劑。

9

將荷葉邊放在上段裙片上，暫時固定。

This is a tiered skirt with two rectangular patterns.
After finishing the gathering, the irons and temporary fixings will make the work beautiful.

1. Arrange the paper templates on the fabric and cut all the sections, then apply fray stopper liquid to all the edges.
2. Fold the seam allowance of the frill hem with an iron.   *3.* Sew the edge.   *4.* Fold the seam allowance of the frill upper with an iron.
5. Make the stitch length 2.5mm and sew one lines on the seam allowance. [refer to p.100 for gathering]
6. Gather to match the width of fit the skirt hem.   *7.* Iron flat.   *8.* Put fabric glue on the hem.   *9.* Place the frill on the hem.

10

縫合荷葉邊。

11

縫合完成後的狀態。

12

將裙子的腰部位置用熨斗熨燙成三折邊。

13

三折邊完成後，車縫壓線。

14

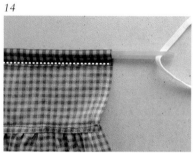

以穿帶器穿過鬆緊帶。

15

將鬆緊帶的一端以珠針固定，由一側完成線
到另一側完成線為止拉出裙子腰部的碎褶
〔S:7cm M:8cm L:8cm〕。

16

以正面相對的方式對齊後縫合。

17

用熨斗將縫份由中間燙開。

18

翻回正面後就完成了。

10. Sew the frill.　*11.* The frill is attached.　*12.* Fold the waist seam allowance 2 times with an iron.
13. Sew the folded.　*14.* String elastic through the waist.　*15.* Gather the waist to [S:7cm　M:8cm　L:8cm] to the finishing line.
16. Sew the back opening together.　*17.* Iron open the seam allowance.　*18.* Turn right side out.

Apron Dress
連衫圍裙

這是可以和罩衫或是連身裙重疊穿搭，附有前擋片的連衫圍裙。
如果在平織薄棉布料加上細褶或是抽花繡裝飾，
看起來就會有如同重疊穿搭般的的加分效果。

平織薄棉布 〔衣身&裙子〕	S 12×44cm M 14×45cm L 15×49cm	平織薄棉布 〔裡布〕	S 10×8cm M 11×8cm L 14×8cm
平織薄棉布 〔腰帶〕	S 3×13cm M 4×15cm L 4×15cm	暗釦	S,M,L 1組

1

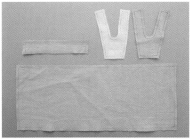

依照紙型將各部位的布片裁切下來，布端塗抹防綻液進行處理。
→「衣身」的應用設計請參考第79-83頁。

2

將衣身的表布與裡布以正面相對的方式對齊。

3

保留上下側的返口，如照片般縫合。

4

在彎弧轉角的縫份剪出細密的牙口。請小心不要剪到縫線。

5

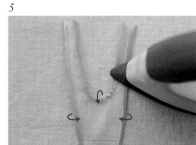

用熨斗將縫份折向裡布那側。

6

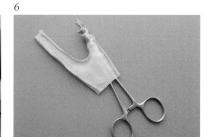

使用鉗子等工具，將布料翻回正面。

7

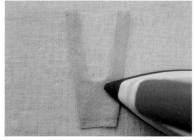

用熨斗燙平。

8

車縫一道壓線。

9

將裙子下襬及兩端的縫份用熨斗燙折至內側。
→「下襬」的應用設計請參考第96-99頁。

The Apron dress can be layered on a blouse or dress.
If you put pin tucks or drawn work with a thin cotton loan, it will look great on a layering and it is wonderful.

1. Arrange the paper templates on the fabric and cut all the sections, then apply fray stopper liquid to all the edges.
[refer to p.79-83 for the front arrangements]   *2.* Match the right sides of the bodice and lining.   *3.* Sew as pictured.
4. Snip the seam allowance. Be careful not to cut the stitches.   *5.* Fold and iron the seam allowance.   *6.* Turn right side out using a tailor's awl and iron.
7. Iron to shape.   *8.* Sew the edges.   *9.* [refer to p.96-99 for the hem arrangements] Fold the seam allowance of the hem and both sides with an iron.

10

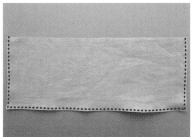

在下襬及兩端車縫一道壓線。

11

在腰部的縫份縫上2道針腳寬度約2.5mm的
抽褶用縫線。
→關於碎褶，請參考第100頁。

12

拉動上側縫線做抽褶。

13

配合腰帶的幅寬，抽出碎褶，然後再用熨斗
整燙固定。

14

將衣身與腰帶以正面相對的方式對齊。此時
腰帶兩端的縫份會比裙片再多出一些。

15

將腰部縫合起來。

16

將縫份倒向腰帶那側，用熨斗燙壓固定。

17

將腰帶兩端的縫份也用熨斗折向內側。

18

將腰帶上部的縫份也用熨斗折向內側。

10. Sew the edges.　11. Use a machine to sew gathering stitches in the upper seam allowance of skirt. [refer to p.100 for gathering]
Make the stitch length 2.5mm and sew two lines on the seam allowance.　12. Gather the fabric to match the width of the bodice waist.
13. Iron flat.　14. With right sides facing, sew the waist belt and the skirt waist together.　15. Sew the waist.
16. Fold the seam allowance toward the waist belt using an iron.　17. Fold the seam allowance of the both sides.
18. Fold the upper seam allowance of the waist belt.

19

在裙子的碎褶部分塗上布用接著劑，將腰帶對折成一半後暫時固定起來。

20

將正面也熨燙平整。

21

在腰帶的正中央、後衣身的兩端縫合位置加上記號。

22

將衣身以布用接著劑暫時固定。

23

在腰帶車縫壓線。

24

在腰帶的周圍車縫了一圈壓線。

25

在腰帶裝上暗釦，這樣就完成了。

19. Fold the waist belt in half with fabric glue.   *20*. Please check it right side.
21. Mark in the waist belt as pictured.   *22*. Create the bodice to the waist belt with fabric glue.
23. Sew the waist belt.   *24*. The edge stitches are finished now.   *25*. Add snaps to the back waist belt to complete.

Knickerbockers

荷葉邊燈籠褲

這是褲管下襬向內收縮，外形輪廓隆起飽滿的六分褲。

沒有荷葉邊的款式，給人簡潔清爽的感覺；有荷葉邊的話則給人可愛的感覺。

麻布/平織薄棉布　S　18×30cm　　　暗釦　S,M,L　1組
　　　　　　　　M　20×30cm
　　　　　　　　L　23×30cm

1

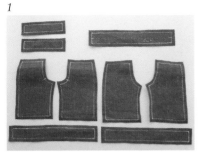

依照紙型將各部位的布片裁切下來，布端塗抹防綻液進行處理。

2

將褲子前股上的正面相對合攏，縫合起來。

3

在彎弧轉角的縫份剪出細密的牙口。請小心不要剪到縫線。

4

將縫份用熨斗燙開。

5

將前褲片和後褲片的兩脇邊正面相對合攏後，縫合起來。

6

將縫份用熨斗燙開。

7

將腰帶正面相對合攏後，縫合起來。

8

用熨斗將縫份倒向腰帶那一側。

9

將腰帶的縫份用熨斗燙折起來。

Plummeted silhouette pants with a squeezed hem.
If there is no frills, it will be neat and pretty style.

1. Arrange the paper templates on the fabric and cut all the sections, then apply fray stopper liquid to all the edges.
2. With the right sides of the left and right front section facing, sew the front rise.   *3.* Cut slits in the seam allowances where the fabric curves.
4. Iron open the seam allowance.   *5.* With the right sides of the front and back facing, sew the sides.   *6.* Iron open the seam allowance.
7. With the right sides of the trousers and waist belt.   *8.* Fold the seam allowance to the waist belt with an iron.
9. Fold the seam allowance of the waist belt with an iron.

10

在腰帶車縫壓線。

11

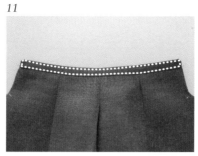

車縫壓線完成後的狀態。

12

在褲子的下襬縫上一道針腳寬度約2.5mm
的抽褶用縫線。
→關於碎褶，請參考第100頁。

13

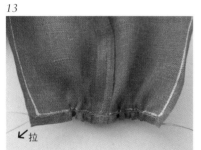

配合褲口布的寬度，拉動縫線抽褶。

14

將碎褶整理好後，用熨斗按壓固定。

15

與褲口布正面相對合攏後，縫合起來。

16

將另一側的褲口下襬與褲口布也一樣縫合起來。

17

將縫份用熨斗倒向袖口布那側。

18

將袖口布下部的縫份用熨斗燙折起來。

10. Sew the waist belt.   11. The waist belt is finished to sew reinforced stitches.
12. Use a machine to sew one line of gathering stitch length 2.5mm on the seam allowance. [refer to p.100 for gathering]
13. Gather to match the width of fit the trousers cuffs.   14. Iron flat.   15. With right sides of the cuffs and trousers.
16. Sew the cuffs.   17. Fold the seam allowance toward to cuffs with an iron.   18. Fold the seam allowance of the under cuffs with an iron.

19

將荷葉邊布片的下襬縫份用熨斗燙折起來。

20

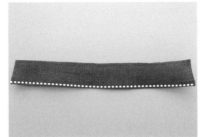

車縫壓線。

21

在上部的縫份縫上一道針腳寬度約2.5mm
的抽褶用縫線。
→關於碎褶，請參考第100頁。

22

拉

配合褲口布的寬度，拉動縫線抽褶。

23

將碎褶整理好後，用熨斗按壓固定。

24

在步驟18折起的褲口布下襬那側的縫份，
塗上布用接著劑。

25

將荷葉邊暫時固定住。

26

確認一下暫時固定後，由正面觀看時的外觀
是否美觀。

27

在褲口布的部分車縫壓線。

19. Fold the seam allowance of the frill hem with an iron.　*20.* Sew the hem.
21. Use a machine to sew one line of gathering stitch length 2.5mm on the upper seam allowance.　*22.* Gather to match the width of fit the cuffs..
23. Iron flat.　*24.* Put fabric glue on the seam allowance of the cuffs hem.　*25.* Attach the frills on the cuffs.
26. Make sure the frills is fine place.　*27.* Sew the cuffs.

28

縫合完成後的狀態。

29

將後褲片的襠部到後開口的記號為止，以正面相對合攏後縫合起來。

30

在彎弧轉角的縫份剪出細密的牙口。

31

用熨斗將縫份燙開。

32

將後開口的縫份用熨斗燙折起來。

33

在後開口車縫壓線。

34

將股下以正面相對合攏後縫合起來。

35

在股下的縫份剪出一道牙口。

36

翻回正面，裝上暗釦之後就完成了。

28. The cuffs are finished stitches.   29. With the right sides of the back facing, sew the back rise to the opening marker.
30. Cut slits in the seam allowances where the fabric curves.   31. Iron open the seam allowance.
32. Fold the seam allowance of the back opening with an iron.   33. Sew the back opening.   34. Sew the inseams together.
35. Cut slits into the seam allowance.   36. Turn the right side out. Add snaps to the back opening.

Jacket
夾克外套

這是以寬鬆的燈籠袖與圓形衣領為款式特徵的夾克外套。

不管是用緞帶輕輕繫在領口，或是用暗釦固定前開口，再加上裝飾鈕釦都很好看。

麻布/天鵝絨	S 18×45cm	(在前開口安裝緞帶時)
	M 20×50cm	絲綢緞帶　　　S、M、L 10cm×2條
	L 22×55cm	〔3.5mm 幅寬〕
荷葉邊緞帶	S 14cm	(在前開口安裝暗釦時)
〔衣領〕	M 15cm	暗釦　　　　　S、M、L 3組
	L 16cm	

1

依照紙型將各部位的布片裁切下來，布端塗抹防綻液進行處理。

2

將2片衣領片正面相對合攏後，沿著外側的完成線縫合起來。

3

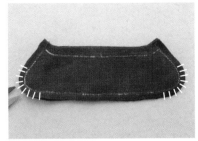

在彎弧轉角的縫份剪出細密的牙口。請小心不要剪到縫線。

4

翻回正面，以錐子或是鉗子整理邊角和彎弧形狀。

5

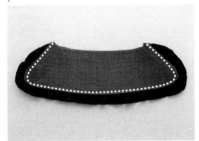

以布用接著劑將荷葉邊緞帶暫時固定。

6

從正面將荷葉邊緞帶縫合起來。

7

縫合完成的狀態。

8

將前衣身與後衣身以正面相對合攏後，將肩部縫合。

9

將縫份用熨斗燙開。

This jacket has a loose puff sleeve and a rounded collar. It is fastened with a ribbon loosely or it is good to close the front opening with a snap and attach a decorative buttons.

1. Arrange the paper templates on the fabric and cut all the sections, then apply fray stopper liquid to all the edges.
2. Take the collar pieces and match the edges, sew.   3. Cut slits in the seam allowance on the round.
4. Turn the piece the right side out, using tailor's awl to neatly push out the corners and curves. Iron to shape.
5. Temporarily fix the frill ribbon with fabric glue.   6. Sew the edges.   7. The frill ribbon is now attached to the collar.
8. Match the front and back of the bodice by the shoulders and sew.   9. Iron open the seam allowance.

10

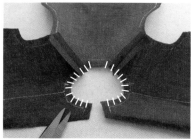

在頸部周圍的縫份剪出細密的牙口。

11

用珠針將衣領及衣身固定住。

12

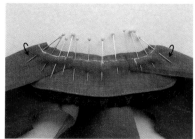

將貼邊布以正面相對的方式折起後,用珠針
固定住。

13

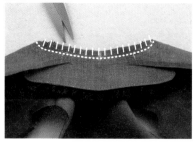

沿著完成線縫合起來。在縫份剪出細密的牙
口,並將邊角修掉。

14

將貼邊翻回正面,用熨斗整平。

15

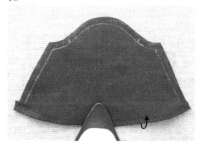

將袖口的縫份用熨斗燙折起來。

16

縫上一道針腳寬度約2.5mm的抽褶用縫線。
→關於碎褶,請參考第100頁。

17

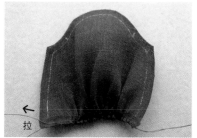

由一側完成線到另一側完成線為止拉出碎褶
〔S:4cm M:4.5cm L:4.5cm〕。

18

將碎褶的間隔整理平均,再用熨斗按壓固定。

10. Cut slit in the seam allowance of the neckline.   *11.* Pin with right sides facing.   *12.* Fold the front facing so it faces inwards.
13. Sew the neckline. Cut fine slits in the seam allowance of the neckline.   *14.* Turn the front facing the right side out, iron to shape.
15. Fold the seam allowance of the sleeve opening inward and iron.   *16.* Machine sew one line of gathering stitches 2.5mm in length in the seam allowance.
[refer to p.100 for gathering]   *17.* Gather to [S:4cm M:4.5cm L:4.5cm] with the finished line.   *18.* Iron flat.

19

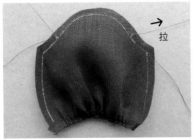

在袖山縫份的兩處記號之間縫上一道針腳寬度約2mm的抽褶用縫線。

20

配合衣身袖襱的幅寬抽碎褶，並將縫線綁好後，用熨斗整燙。

21

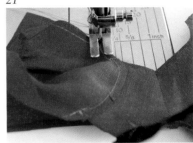

將衣袖與衣身正面相對合攏後，縫合起來。

22

將縫紉機的壓腳不時抬高，確認袖山的縫份與袖襱的縫份隨時保持對齊合攏，一點一點前進縫合起來。

23

衣袖固定在衣身上了。將縫份倒向衣袖那側，再用熨斗整燙。

24

將前衣身與後衣身正面相對合攏後，沿著袖口～脇邊～下襬縫合。

25

在脇邊的縫份剪出牙口。

26

翻回正面，將縫份用熨斗燙開。

27

將貼邊布的下襬正面相對合攏後，以珠針固定。

19. Machine sew one line of gathering stitches 2mm in length in the sleeve cap seam allowance from marker to marker.
20. Gather the shoulders until the width fits the armhole and iron. 21. Match the side edge of the sleeve to the bodice and gradually, sew the shoulder of the sleeve to armhole. 22. Raise the machine presser foot a number of time while sewing to gradually align the seam allowances of the sleeve caps and armholes. 23. Now the sleeves are attached. Place the seam allowance toward the sleeves and iron. 24. With the right sides of the front and back bodice facing, sew them together. 25. Cut slits into the seam allowance of the pits. 26. Turn right side out. Iron open the seam allowance. 27. Pin the hem of the front facing.

28

將貼邊布的下襬縫合起來。

29

修掉縫份的邊角。

30

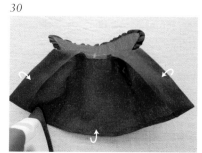

將貼邊布翻回正面,用熨斗整燙。下襬的縫份也用熨斗燙折起來。

31

接著要開始在頸部周圍車縫壓線。

32

沿著前開口～下襬～前開口～頸部周圍車縫一圈壓線。

33

壓線車縫完成後的狀態。

34

在領口縫上一條10cm的緞帶(在前開口裝上一個暗釦也可以)。

35

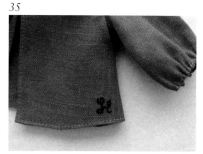

用刺繡線以單線的方式繡上名字英文發音的第一個字母。

36

完成。

28. Sew the hem of the front facing.   29. Cut the corners.   30. Turn right side out of the front facing and iron. Fold the hem.
31-33. Sew around the edges from neckline along the front opening and the hem, then back up to the neckline.
34. Sew the 10cm ribbon to the front opening. Or you can add snaps to the front opening.
35. Take a single embroidery and sew reverse stitches for initial.   36. Complete the jacket.

Coat
大衣

這是百搭萬用的A字大衣。由於衣領的紙型與夾克外套共通的關係，

張開的時候是西裝領，閉合的時候則變身成為可愛的圓領。

可以自由呈現出不同的款式氣氛。

羊毛布/麻布　S　15×48cm　　　　暗釦　S,M,L　4組
　　　　　　　M　17×55cm
　　　　　　　L　22×60cm

荷葉邊緞帶　S　14cm
〔衣領〕　　 M　15cm
　　　　　　 L　16cm

1

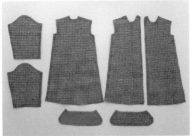

依照紙型將各部位的布片裁切下來，布端塗
抹防綻液進行處理。

2

將2片衣領片正面相對合攏後，沿著外側的
完成線縫合起來。

3

在彎弧轉角的縫份剪出細密的牙口。請小心
不要剪到縫線。

4

翻回正面，以錐子或是鉗子整理邊角和彎弧
形狀。

5

車縫壓線固定。

6

縫合完成的狀態。

7

將前衣身與後衣身正面相對合攏後，肩部縫
合起來。

8

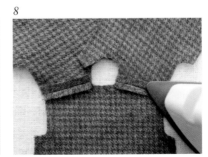

將縫份用熨斗燙開。

9

在頸部周圍的縫份剪出細密的牙口。

*The collar pattern is common with the jacket, When opened, it turns into a tailored collar,
and when closed, it turns into a cute round collar. Enjoy the difference.*

1. Arrange the paper templates on the fabric and cut all the sections, then apply fray stopper liquid to all the edges.
2. Take the collar pieces and match the edges, sew. 3. Cut slits in the seam allowance on the round.
4. Turn the piece the right side out, using tailor's awl to neatly push out the corners and curves. Iron to shape. 5. Sew the edges of the collar.
6. Finished to sew the edges. 7. Match the front and back of the bodice by the shoulders and sew.
8. Iron open the seam allowance. 9. Cut slits in the seam allowance of the neckline.

10

將衣領與衣身以珠針固定。

11

將貼邊布折至正面相對合攏，再以珠針固定。

12

沿著完成線縫合起來。在縫份剪出細密的牙口，並將邊角修掉。

13

將貼邊翻回正面，用熨斗整平。

14

將袖口的縫份用熨斗燙折起來。

15

車縫壓線。

16

在袖山縫份的兩處記號之間縫上一道針腳寬度約2mm的抽褶用縫線。

17

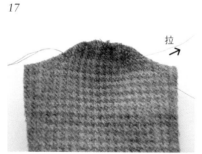

配合衣身袖襱的幅寬抽碎褶，並將縫線綁好。

18

將袖山的碎褶整理好後，用熨斗整燙一遍。

10. Pin with right sides facing. 11. Fold the front facing so it faces inwards. 12. Sew the neckline. Cut fine slits in the seam allowance of the neckline.
13. Turn the front facing the right side out, iron to shape. 14. Fold the seam allowance of the sleeve opening inward and iron.
15. Sew the sleeve opening. 16. Machine sew one line of gathering stitches 2.5mm in length in the seam allowance.
17. Gather the shoulders until the width fits the armhole. 18. Iron flat.

19

將衣袖及衣身正面相對合攏後縫合起來。

20

將縫紉機的壓腳不時抬高,確認袖山的縫份與袖襱的縫份隨時保持對齊合攏,一點一點前進縫合起來。

21

慢慢向前推進,就能縫出漂亮的縫線。

22

將衣袖固定在衣身上了。

23

將縫份倒向衣袖那側,再用熨斗整燙。然後將前衣身和後衣身正面相對合攏對齊。

24

沿著袖口~脇邊~下襬縫合起來。

25

在脇邊的縫份剪出一道牙口。

26

翻回正面,將縫份用熨斗燙開。

27

將貼邊布的下襬正面相對合攏後,用珠針固定。

19-22. Match the side edge of the sleeve to the bodice and gradually, sew the shoulder of the sleeve to armhole.
Raise the machine presser foot a number of times while sewing to gradually align the seam allowances of the sleeve caps and armholes.
23. Place the seam allowance toward the sleeves and iron. With the right sides of the front and back bodice facing.
24. Sew them together.   *25.* Cut slits into the seam allowance of the pits.
26. Turn right side out. Iron open the seam allowance.   *27.* Pin the hem of the front facing.

28

將貼邊布的下襬縫合起來，修掉縫份的邊角。

29

將貼邊翻回正面，下襬的縫份用熨斗燙折起來。

30

由頸部周圍開始車縫壓線。

31

由頸部周圍朝向前開口。

32

由前開口朝向下襬。

33

沿著頸部周圍～前開口～下襬～前開口～頸部周圍車縫一圈壓線後的狀態。

34

用熨斗整理衣領的形狀。

35

最後裝上暗釦和珠子就完成了。

36

將衣領形狀改為不是西裝領的應用設計也很可愛。

28. Sew the hem. Cut the corners.   29. Turn right side out of the front facing and iron. Fold the hem.
30-33. Sew around the edges from neckline along the front opening and the hem, then back up to the neckline.
34. Iron to shape.   35. Add snaps and beads.   36. It is also cute not to make the shape of the collar tailored.

Hat
帽子

這是平均的1/6尺寸娃娃,以及Blythe這種頭形較大的娃娃戴的帽子。
可以纏繞緞帶、加上花朵裝飾等等,請自由發揮各種設計吧!

麻布	1/6尺寸	12×30cm
	Blythe尺寸	35×50cm

1

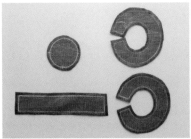

依照紙型將各部位的布片裁切下來，布端塗抹防綻液進行處理。
→Blythe尺寸請參考第72頁。

2

將側邊帽片正面相對合攏。

3

縫合起來。

4

將縫份用熨斗燙開。

5

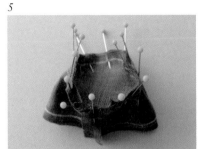

將頂部帽片與側邊帽片正面相對合攏對齊。

6

如果不容易機縫的話，可以取雙線以手縫半回針的方式縫合。

7

縫合完成的狀態。

8

在縫份剪出細密的牙口。請小心不要剪到縫線。

9

翻回正面。

A hat for the average 1/6 doll and a doll with a big head such as Blythe.
Please enjoy wrapping ribbons and decorating flowers.

1. Arrange the paper templates on the fabric and cut all the sections, then apply fray stopper liquid to all the edges. [Blythe size p.72]
2. Match the side crown together inside out.   *3.* Sew the side crown.   *4.* Iron open.   *5.* Match the top crown and side crown together inside out.
6-7. If it is difficult to sew with sewing machine, you can sew by hand with two threads to back stitch.
8. Cut slits the seam allowance. Be careful not to cut the stitches.   *9.* Turn right side out.

10

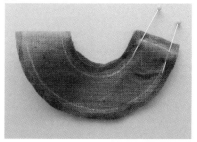

將帽緣布片正面相對合攏對齊。

11

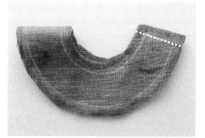

縫合起來。

12

用熨斗將縫份燙開。

13

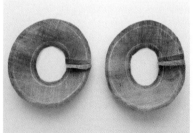

將表裡 2 塊帽緣布片縫合完成後的狀態。

14

以正面相對的方式合攏對齊，沿著外側的完成線縫合起來。

15

在縫份剪出細密的牙口。請小心不要剪到縫線。

16

翻回正面，用熨斗將形狀整理成漂亮的圓形。

17

在外側車縫壓線。

18

將帽緣布片與帽身正面相對合攏後，用珠針固定住。

10. Match the brim together inside out.   *11.* Sew together.   *12.* Iron open.   *13.* Sew two pieces.
14. Match the brim inside out and sew.   *15.* Cut slits the seam allowance.   *16.* Turn right side out. Iron to shape.
17. Sew the edges.   *18.* Pin the brim and crown inside out together.

19

如果不容易機縫的話，可以取雙線以手縫半回針的方式縫合。

20

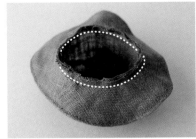

縫合完成的狀態。

21

在縫份剪出細密的牙口。

22

用熨斗將縫份倒向帽身那側後熨燙整平，完成。

23

纏繞上蕾絲的應用設計。

24

以平針縫將數種不同的蕾絲固定在帽子上的應用設計。

25

將細緞帶纏繞上去打一個蝴蝶結的應用設計。

19-20. If it is difficult to sew with sewing machine, you can sew by hand with two threads to back stitch.
22. Fold the seam allowance toward to the crown with an iron.   *23.* This is arrangement with laces.
24. Add the several laces are together with running stitch.   *25.* This is arrangement with thin ribbon.

1

頭形較大的娃娃尺寸用的帽子要加上裡布。與表布相同，先沿著紙型將各布片裁剪下來，再將布端塗抹上防綻液進行處理。

2

將側邊帽片以正面相對的方式圍成一圈，把縫份燙開。然後再與頂部帽片正面相對合攏對齊後，用珠針固定住。

3

將頂部帽片與側邊帽片縫合起來。

4

在縫份剪出細密的牙口。請小心不要剪到縫線。

5

用熨斗將縫份燙開。裡布也是相同的作業方式。

6

將裡布裝入表布的內側。

7

將帽緣依前頁步驟10～17進行縫合，然後與帽身以正面相對的方式合攏對齊，用珠針固定。

8
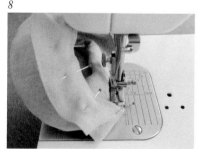

將帽緣與帽身縫合起來。

9

在縫份上剪出細密的牙口，然後用熨斗將縫份倒向帽身那側整平後就完成了。

Huge Hat

1. Arrange the paper templates on the fabric and cut all the sections, then apply fray stopper liquid to all the edges. The huge hat needs back fabric.
2. Match the side crown together inside out and sew. Iron open. Match the top crown and side crown together inside out.
3. Sew together.   *4.* Cut slits the seam allowance. Be careful not to cut the stitches.   *5.* Iron open. Make the back fabric in the same way.
6. Set the back fabric inside the crown of the right side.   *7.* Sew the brim. [Refer to 10-17.] Pin the crown and the brim together.   *8.* Sew together.
9. Cut slits the seam allowance. Turn right side out. Fold the seam allowance toward to the crown with an iron.

Bag
包包

這裡是使用皮革、麂皮這類不需要處理布端的合成皮革製作包包。
建議使用厚度較薄的面料,如果擔心強度不足的話,可以將布端縫線壓邊。

合成皮革　各 8×10cm

刺繡線　　依個人喜好

1

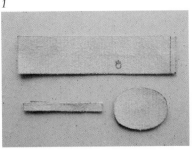

依照直筒型手提包的紙型,將各部位的布片裁切下來。

2

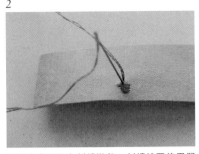

視個人喜好加上刺繡裝飾。刺繡線要使用雙線縫法。

3

草莓的刺繡完成了。

4

在縫份塗抹皮革用接著劑。

5

將端面重疊貼合,圍成圓筒。

6

在底面布片的邊緣塗上接著劑。

7

將圓筒和底面組合黏貼起來。

8

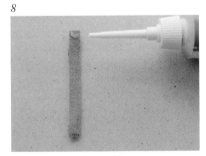

在提把的兩端塗抹接著劑。

9

將提把安裝上去就完成了。

For the bag, use leather, suede, or synthetic leather. It is recommended to use thin materials.
If the strength is insecure, you can sew the edges.

1. Arrange the paper templates on the fabric and cut all the sections.   2. Use two embroidery threads.   3. The strawberry embroidery is finished.
4. Put leather glue on the edges.   5. Match the edges together and make a tube.   6. Put leather glue on the edges.
7. Match together.   8. Put leather glue on the edges of the handle.   9. Paste together.

1

依照大手提包的紙型，將各部位的布片裁切下來。

2

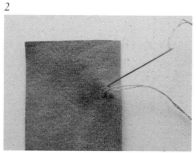

視個人喜好加上刺繡裝飾。刺繡線要使用雙線縫法。

3

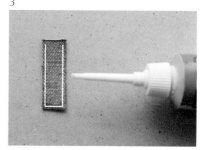

在側邊布片的邊緣塗抹皮革用接著劑。

4

將包包本體沿著完成線折疊起來，然後將側邊布片黏貼組合上去。接下來再將底部與側邊折出折痕。

5

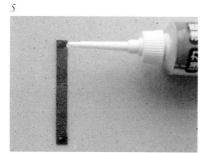

在提把的兩端塗抹接著劑。

6

將提把安裝上去就完成了。

7

可以改變提把的長度、包包本體的大小，做出各種不同的應用設計。

1. Arrange the paper templates on the fabric and cut all the sections.　*2*. Use two embroidery threads.　*3*. Put leather glue on the edges.
4. Fold the bag and paste together with side pieces. And fold the bottom and the sides.　*5*. Put leather glue on the edges.　*6*. Paste together.
7. You can make various arrangements to change the length of the handle or the size of the bag.

Front
衣身的應用設計

Collar
衣領的應用設計

Sleeve
衣袖的應用設計

Hem
下襬的應用設計

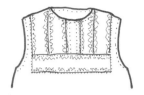

a.
Lace Collage
蕾絲拼貼

這是在前衣身的應用設計中最簡單的蕾絲拼貼，只要將蕾絲放上去縫合起來就完成了。

1

將自己喜歡的蕾絲尺寸放置於前衣身的中心。

2

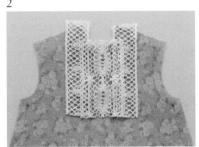

兩側也放上蕾絲。

3

將蕾絲稍微重疊擺放，調整成左右對稱的狀態。

4

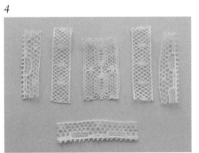

先將蕾絲自衣身取下。

5

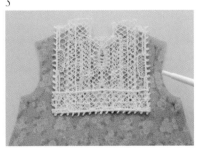

以布用接著劑將蕾絲一片一片暫時固定住。

6

縫合起來。

7

縫合完成後的狀態。

8

配合衣身的外形，修剪蕾絲。

9

就算只是縫上數條直向的蕾絲也很可愛。

1. Place your favorite size lace in the center of the front bodice. *2.* Place the lace in the both of sides.
3. Place the laces little by little symmetrically. *4.* Take off the laces from the front bodice.
5. Apply fabric glue the laces. *6-7.* Sew the laces. *8.* Cut the laces. *9.* It is cute just sewing several laces vertically.

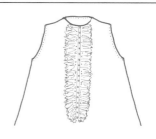

b.

Lace Frill

蕾絲荷葉邊

如果有多餘的蕾絲，那就抽褶製作成荷葉邊裝飾吧！

1

在前衣身的中心畫上一道自己喜歡長度的線條。

2

準備一條比畫在前衣身線條長2倍的蕾絲，並在正中央畫線。

3

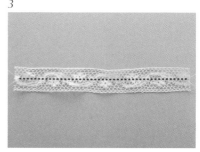

縫上一道針腳寬度約2mm的抽褶用縫線。
→關於碎褶，請參考第100頁。

4

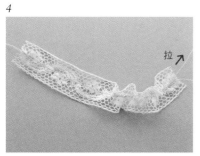

拉↗

配合前衣身線條的長度抽碎褶，並將縫線綁好。

5

整理好碎褶的形狀，用熨斗按壓固定。

6

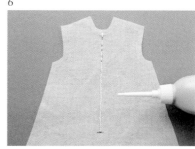

將布用接著劑塗抹在前衣身的線條上。

7

將蕾絲暫時固定上去。

8

將一條細緞帶暫時固定在針腳上，然後縫合起來。

9

最後縫上珠子或是鈕釦作為裝飾也很可愛。

1. Draw the line of desired length in the center of the front bodice. *2.* Prepare a lace that is about twice as long as the line. Draw the line in the center of the lace. *3.* Machine sew one line of gathering stitches 2mm in length in the center. [refer to p.100 for gathering] *4.* Gather the lace until the width fits the line. *5.* Iron flat. *6.* Apply fabric glue on the line. *7.* Temporarily the lace. *8.* Apply fabric glue on the center of the lace. Place the thin ribbon and sew. *9.* It is cute for decoration with beads or buttons for finishing.

c.
Lace Pleats
蕾絲褶襉

將單邊蕾絲製作成褶襉，呈現出華麗的感覺。正中央的細緞帶使用不同顏色，看起來更漂亮。

1

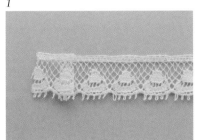

先在蕾絲折出5mm的褶襉。

2

每折一個褶襉就以布用接著劑固定，比較好作業。

3

直到折出想要使用幅寬的褶襉為止。

4

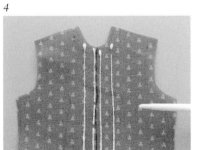

將布用接著劑塗抹在前衣身。

5

放上蕾絲褶襉，暫時固定。

6

正中央放一條細緞帶，暫時固定。

7

縫合起來。

8

縫合完成後的狀態。

9

最後縫上珠子或是鈕釦作為裝飾也很可愛。

1. About 5mm pleats to the lace. *2.* It is easy to make with fabric glue. *3.* Make pleats to the width you want to use.
4. Apply fabric glue on the front bodice. *5.* Temporarily the lace. *6.* Apply fabric glue on the center of the lace. Place the thin ribbon.
7-8. Sew the ribbon. *9.* It is cute for decoration with beads or buttons for finishing.

d.
Pin Tuck
蕾絲細褶

蕾絲細褶非常適合與裝飾假領一起搭配。也可以應用在罩衫和圍裙的設計上。

1

準備一塊比想要使用的紙型尺寸更大的面料。

2

沿著布料的布紋，用熨斗燙折起來。

3

依照自己喜好的幅寬，在折痕的旁邊畫一條平行線（照片是距離折痕1mm）。

4

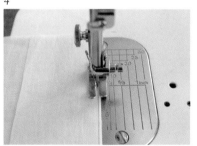

在這條線上進行縫合。

5

縫合完成後的狀態。

6

將面料攤開，把先前縫好的折痕倒向外側，再用熨斗整燙。

7

同樣沿著布料的布紋，用熨斗燙折起來（依自己喜歡的幅寬即可）。

8

再畫上一條縫線，然後縫合起來。

9

將面料攤開，把先前縫好的折痕倒向外側，再用熨斗整燙。

1. Prepare a fabric that is larger than the pattern you want to use.　*2.* Fold along the weave with an iron.
3. Draw a seam line from the crease at the desired width (the photo is 1 mm from the crease).　*4-5.* Sew the line.
6. Iron open and iron toward to outside.　*7.* Fold in the same way along the texture.　*8.* Draw the line and sew.　*9-10.* Iron open and outward.

10

細褶縫合完成後的狀態。

11

將紙型複寫到面料上。

12

裁剪下來。

13

將面料的端面塗抹防綻液進行處理。

14
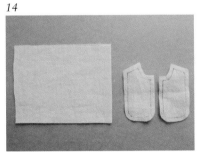
如果是裝飾假領搭配細褶的應用設計，先將想要加上細褶的布片零件裁剪得更大一些。

15

將細褶製作出來。

16
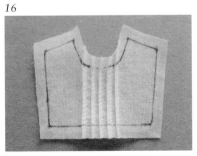
依照紙型將各部位的布片裁切下來，布端塗抹防綻液進行處理。

17
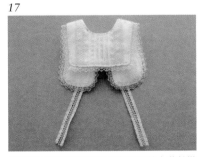
也可以讓裝飾假領在細褶的兩側縫上蕾絲裝飾。

11. Trace the pattern on the fabric. 12. Cut the section. 13. Apply fray stopper liquid to the edges.
14. For detachable collar, Cut the parts you want to pin-tack large. 15. Make pin tucks. 16. Arrange the paper templates on the fabric and cut the section, then apply fray stopper liquid to all the edges. 17. This detachable collar has laces on the sides of the pin tucks.

e.

Ruffled Collar
荷葉邊領

分量感十足，可以瞬間增加華麗氣氛的荷葉邊領。使用較寬的蕾絲製作也很漂亮哦。

1

將荷葉邊布片的縫份用熨斗燙折起來。

2

車縫壓線。

3

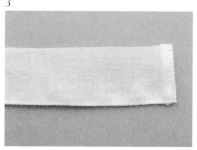

壓線的特寫。

4

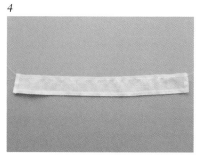

在縫份縫上一道針腳寬度約2mm的抽褶用縫線。

5

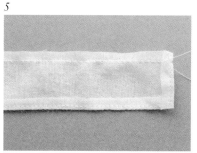

在完成線稍為上方的位置縫上一道抽褶用的縫線。
→關於碎褶，請參考第100頁。

6

配合衣身的袖襱幅寬抽碎褶，並將縫線綁好。

7

將碎褶的間隔整理平均，再用熨斗按壓固定。

8

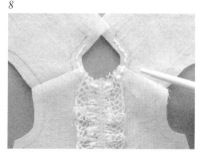

在頸部周圍的縫份塗上布用接著劑。

9

將衣領暫時固定住。

1. Fold the seam allowance of the ruffled with an iron.   *2-3.* Sew the edges.
4. Use a machine to sew one line of gathering stitch length 2mm on the upper seam allowance.
5. Sew gathering stitch slightly above the finished line. [refer to p.100 for gathering]   *6.* Gather the fabric to match of the width of the neckline.
7. Neaten the spacing of the gathering and iron flat.   *8.* Apply fabric glue to the seam allowance.   *9.* Place the ruffled collar.

10

在後衣身貼邊布的縫份塗上布用接著劑。

11

將貼邊布以正面相對合攏，與衣領對齊後，暫時固定。

12

將完成線畫出來。

13

縫合頸部周圍。

14

在縫份剪出細密的牙口，並將邊角修掉。

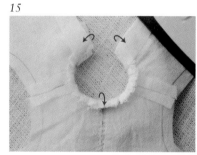

15

將貼邊布翻回正面，用熨斗將衣領的縫份倒向裡側。

16

沿著後開口～頸部周圍～後開口，車縫壓線固定。

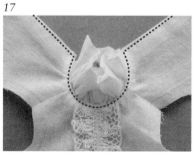

17

車縫壓線完成的狀態。

18

用噴霧器將衣領用水噴濕，再用熨斗整燙後就完成了。

10. Apply fabric glue to the seam allowance of the back facing. *11.* Match the back facing. *12.* Trace the finishing line. *13.* Sew the neck line. *14.* Cut slits in the seam allowances and cut the corners. *15.* Turn right side out of the back facing. Fold inside the seam allowance of the collar. *16-17.* Sew the edges as pictured. *18.* Iron with little water into shape.

f.
Peter Pan Collar
圓領

這是領圍的彎弧角度和緩，很容易縫製的圓領款式。
如果使用厚面料製作的話，建議在裡布使用薄面料。

1

將一對衣領的紙型複寫在面料上，然後裁切成較大的尺寸。另外準備一塊相同大小的面料。

2

將兩塊面料正面相對合攏後，沿著外側的完成線縫合。

3

保留縫份，裁切下來，並將邊角修掉。

4

在彎弧轉角的縫份剪出細密的牙口。請小心不要剪到縫線。

5

翻回正面，以錐子或是鉗子整理邊角和彎弧形狀，然後用熨斗整燙。

6

在縫份塗抹防綻液。

7

在衣身的頸部周圍的縫份剪出細密的牙口。

8

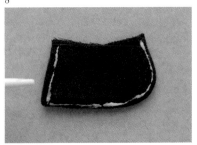

如果要在衣領加上蕾絲的話，將布用接著劑塗抹在衣領背面。

9

加上蕾絲暫時固定。

1. For the collar take two pieces of the same size and draw the collar on one piece. *2.* Take the collar pieces and match the edge, sew along the outer seam line. *3.* Cut the collar sections out, leaving seam allowance, and cut the corners. *4.* Cut slits in the corner. Be careful not to cut the stitches. *5.* Turn right side out with tailor's awl then iron into shape. *6.* Apply fray stopper liquid to the seam allowance. *7.* Cut slits in the seam allowance. *8.* If you want to put the lace around, then apply fabric glue on the edge of the back collar. *9.* Place the laces.

10

車縫壓線。

11

調整至左右對稱後，將衣領暫時固定上去。

12

在衣領上畫出完成線，並在後衣身的貼邊布縫份塗上布用接著劑。

13

將貼邊布以正面相對的方式合攏，與衣領重疊後暫時固定起來。

14

將衣領縫合固定。在衣領的縫份剪出細密的牙口，並將邊角修掉。

15

將貼邊布翻回正面，用熨斗將衣領的縫份倒向裡側。

16

沿著後開口～頸部周圍～後開口車縫壓線。

17

壓線車縫完成後的狀態。

18

圓領完成了。

10. Sew the edges. *11.* Make sure to be symmetric, apply fabric glue on the seam allowance of the neck and place the collars.
12. Trace the finishing line and apply fabric glue on the seam allowance of the back facing. *13.* Fold the back facing.
14. Sew the collars, cut slits in the seam allowance and corners. *15.* Turn right side out of the back facing. Fold inside the seam allowance of the collar.
16-17. Sew the edges as pictured. *18.* Iron into shape.

g.
Lace Stand Collar
蕾絲領

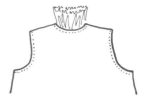

這是用較薄的蕾絲花邊製作的簡單立領，不論長或短都很好看。

1

準備〔S:11cm M:12cm L:15cm〕長度的
蕾絲花邊布。

2

在蕾絲的上部縫上一道針腳寬度約2mm的
抽褶用縫線。
→關於碎褶，請參考第100頁。

3

這是抽褶用縫線的特寫鏡頭。

4

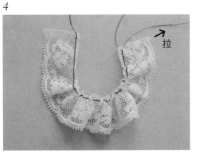

拉

將其中一側的縫線綁好，然後拉動另一側的
下縫線抽出碎褶。

5

符合頸圍尺寸製作出所需碎褶長度後，綁好
縫線，然後再用熨斗整燙。

6

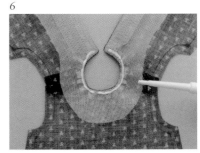

在放上裡布狀態的衣身頸部周圍塗抹布用接
著劑。

7

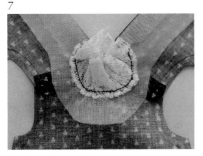

將蕾絲布暫時固定住。

8

沿著後開口～頸部周圍～後開口，車縫壓線
固定。

9

蕾絲領完成了。

---

1. The lace for the collar width is [S:11cm M:12cm L:15cm].
2-3. Use a machine to sew one line of gathering stitch length 2 mm on the upper seam allowance. [refer to p.100 for gathering]
4. Tie the ends at the one side and pull the bottom thread from one side.   *5.* Match the width of fit the neck line and tie the threads.
6. Apply fabric glue on the seam allowance.   *7.* Place the lace collar.   *8-9.* Sew the edges as pictured.

h.
Lace Cap Sleeve
蕾絲袖

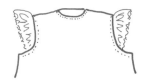

第一次嘗試製作衣袖時，建議使用蕾絲袖。左右的幅寬相同，外形輪廓看起來就很漂亮。

1

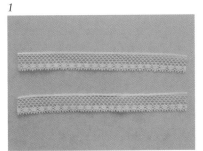

準備2條〔S:8cm M:9cm L:10cm〕長度的蕾絲花邊布。建議選用1～1.5cm幅寬的蕾絲。

2

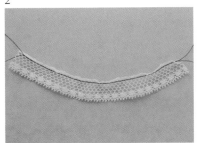

在蕾絲的上部縫上一道針腳寬度約2mm的抽褶用縫線。

3

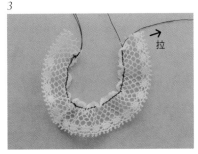

拉

將其中一側的縫線綁好，然後拉動另一側的下縫線抽出碎褶。

4

將蕾絲布抽出〔S:3.5cm M:4cm L:4.5cm〕長度的碎褶，然後把縫線綁好。

5

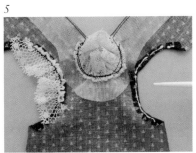

在袖襱的縫份剪出細密的牙口，折進內側，再以布用接著劑將蕾絲暫時固定上去。

6

車縫壓線。

7

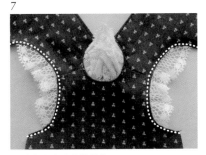

壓線車縫完成後的狀態。

8

將衣身正面相對合攏後，縫合兩側脇邊。

9

翻回正面後就完成了。

1. The lace for the collar width is [S:8cm M:9cm L:10cm].   *2.* Use a machine to sew one line of gathering stitch length 2 mm on the upper seam allowance.
3. Tie the ends at the one side and pull the bottom thread from one side.   *4.* Match the width [S:3.5cm M:4cm L:4.5cm] and tie the threads.
5. Cut fine slits into the seam allowance of the armholes. Fold the seam allowances and temporarily the lace with fabric glue.
6-7. Sew the armholes.   *8.* With right sides of front and back facing, sew together.   *9.* Turn right side out.

i.
Set-in Sleeve
長袖

長袖講究的重點在於外形輪廓，藉由袖山的小碎褶來呈現出立體感。

1

依照紙型將各部位的布片裁切下來，布端塗抹防綻液進行處理。

2

將袖口的縫份用熨斗燙折起來。

3

如果想要在袖口加上蕾絲的話，這裡要塗上布用接著劑。

4

將蕾絲暫時固定起來。

5

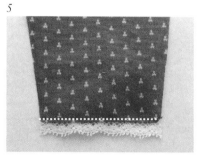

車縫壓線。

6

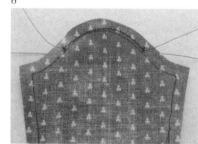

在袖山縫份的兩處記號之間縫上一道針腳寬度約2mm的抽褶用縫線。

7

配合衣身的袖襱幅寬抽出碎褶，然後將縫線綁好。

8

將碎褶整理好後，用熨斗按壓固定。

9

將衣身與衣袖正面相對合攏後縫合起來。

1. Arrange the paper templates on the fabric and cut all the sections, then apply fray stopper liquid to all the edges.
2. Fold the seam allowance of the sleeve opening. 3. If you want to put the lace on the sleeve opening, apply fabric glue.
4. Temporarily the lace with fabric glue. 5. Sew the sleeve opening. 6. Machine sew one line of gathering stitches 2mm in length in the seam allowance.
7. Gather the shoulders until the width fits the armhole. 8. Iron flat.

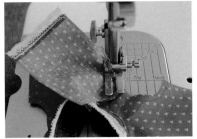

10

將縫紉機的壓腳不時抬高，確認袖山的縫份與袖襱的縫份隨時保持對齊合攏，一點一點前進縫合起來。

11

慢慢向前推進，就能縫出漂亮的縫線。

12

將衣袖固定在衣身上了。

13

將縫份倒向衣袖那側，然後將前衣身和後衣身正面相對合攏對齊，沿著袖口～脇邊～下襬縫合起來。

14

在脇邊的縫份剪出一道牙口。

15

翻回正面，將縫份用熨斗燙開。

16

完成了。

9-12. Match the side edge of the sleeve to the bodice and gradually, sew the shoulder of the sleeve to armhole.
Raise the machine presser foot a number of times while sewing to gradually align the seam allowances of the sleeve caps and armholes.
13. Place the seam allowance toward the sleeves and iron. With the right sides of the front and back bodice facing. Sew them together.
14. Cut slits into the seam allowance of the pits. 15. Turn right side out. Iron open the seam allowance. 16. Iron into shape.

j.
Balloon Sleeve
氣球袖

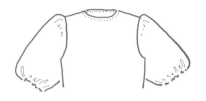

袖山的形狀平緩，一點一點慢慢縫合就能縫出漂亮的外形。

1

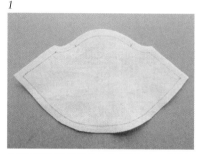

依照紙型將各部位的布片裁切下來，布端塗抹防綻液進行處理。

2

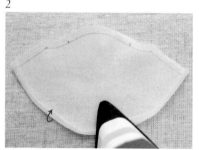

用熨斗將袖口的縫份燙折起來。

3

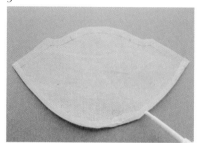

塗抹布用接著劑。

4

將蕾絲暫時固定住。

5

縫上一道針腳寬度約2mm的抽褶用縫線。

6

拉

將其中一側的縫線綁好，然後拉動另一側的上縫線抽出碎褶。

7

在兩側完成線之間將抽出〔S:4cm M:4cm L:4.5cm〕長度的碎褶，然後把縫線綁好。

8

車縫壓線。

9

壓線車縫完成後的狀態。

1. Arrange the paper templates on the fabric and cut all the sections, then apply fray stopper liquid to all the edges.
2. Fold the seam allowance of the sleeve opening.　3. Apply fabric glue.　4. Temporarily the lace with fabric glue.
5. Machine sew one line of gathering stitches 2mm in length in the seam allowance.　6. Tie the ends at the one side and pull the bottom thread from one side.
7. Gather the width [S:4cm M:4cm L:4.5cm] from finished line to finished line. Tie the threads.　8-9. Sew the sleeve opening.

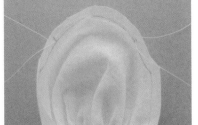

在袖山縫份的兩處記號之間,縫上一道針腳寬度約2mm的抽褶用縫線。

配合衣身袖襱的幅寬抽碎褶,然後把縫線綁好。

整理好碎褶形狀後,用熨斗燙壓固定。

將衣身與衣袖正面相對合攏後縫合。作業中要不時抬高縫紉機的壓腳,確認袖山的縫份與袖襱的縫份隨時保持對齊合攏,一點一點前進縫合起來。

衣袖固定在衣身上了。

將前後衣身正面相對合攏後,沿著袖口~脇邊~下襱縫合。

縫合完成後的狀態。

在脇邊的縫份剪出牙口。

翻回正面,將縫份用熨斗燙開,就完成了。

10. Machine sew one line of gathering stitches 2mm in length in the seam allowance. 11. Gather the shoulders until the width fits the armhole. 12. Iron flat. 13-14. Match the side edge of the sleeve to the bodice and gradually, sew the shoulder of the sleeve to armhole. Raise the machine presser foot a number of times while sewing to gradually align the seam allowances of the sleeve caps and armholes. 15-16. Place the seam allowance toward the sleeves and iron. With the right sides of the front and back bodice facing. Sew them together. 17. Cut slits into the seam allowance of the pits. 18. Turn right side out. Iron open the seam allowance.

k.
Band Cuffs
附袖口布衣袖

和氣球袖相較下分量感沒那麼強烈，也可以不裝上袖口布呈現氣球袖般的感覺。

1

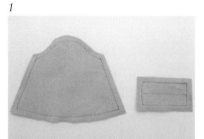

依照紙型將各部位的布片裁切下來，布端塗抹防綻液進行處理。

2

在袖口的縫份縫上一道針腳寬度約2mm的抽褶用縫線。

3

將其中一側的縫線綁好，然後拉動另一側的上縫線抽出碎褶。

拉

4

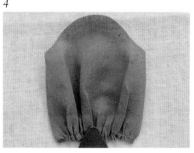

配合袖口布的幅寬抽碎褶，然後把縫線綁好後再用熨斗燙壓固定。

5

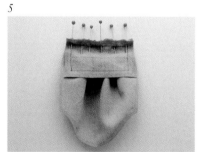

將袖口與袖口布正面相對合攏。

6

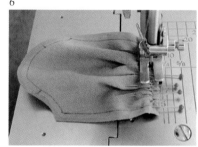

縫合袖口。

7

縫合完成後的狀態。

8

將縫份用熨斗倒向袖口布那側。

9

將袖口布的縫份用熨斗燙折起來。

1. Arrange the paper templates on the fabric and cut all the sections, then apply fray stopper liquid to all the edges.
2. Machine sew one line of gathering stitches 2mm in length in the seam allowance.
3. Gather the sleeve opening until the width fits the cuff. *4.* Iron flat. *5.* Match the sleeve and cuff.
6-7. Sew the sleeve opening. *8.* Fold the seam allowance toward to cuff. *9.* Fold the seam allowance of the cuff with an iron.

10

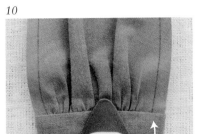

將袖口布折成一半。

11

如果想要加上蕾絲的話，此時要塗上布用接著劑。

12

將蕾絲暫時固定住。

13

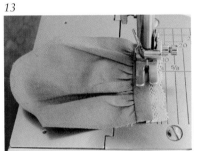

在袖口布車縫壓線。

14

壓線車縫完成後的狀態。

15

↗拉

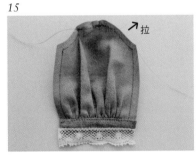

在袖山縫份上的兩處記號之間，縫上一道針腳寬度約2mm的抽褶用縫線。配合衣身袖襱的幅寬抽碎褶，並將縫線綁好。

16

將衣身與衣袖正面相對合攏後縫合。作業中要不時抬高縫紉機的壓腳，確認袖山的縫份與袖襱的縫份隨時保持對齊合攏，一點一點前進縫合起來。

17

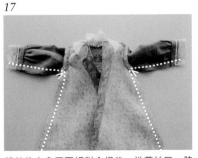

將前後衣身正面相對合攏後，沿著袖口～脇邊～下襬縫合。並在脇邊的縫份剪出一道牙口。

18

翻回正面後就完成了。

10. Fold the seam allowance in half. *11.* If you want to put the lace, apply fabric glue. *12.* Temporarily the lace. *13-14.* Sew the cuffs. *15.* Machine sew one line of gathering stitches 2mm in length in the seam allowance. Gather the shoulders until the width fits the armhole. *16.* Match the side edge of the sleeve to the bodice and gradually, sew the shoulder of the sleeve to armhole. Raise the machine presser foot a number of times while sewing to gradually align the seam allowances of the sleeve caps and armholes. *17.* Place the seam allowance toward the sleeves and iron. With the right sides of the front and back bodice facing. Sew them together. *18.* Cut slits into the seam allowance of the pits. Turn right side out. Iron open the seam allowance.

l.
Lace Gather
蕾絲碎褶

可以縫在下襬的裡側作為裝飾。如果不抽碎褶就直接縫上去的話，看起來相當清爽簡單。

1

丈量想要裝上蕾絲碎褶裝飾的位置，準備一條約2倍長度蕾絲。

2

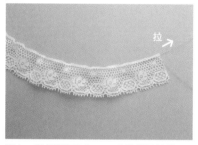

拉 ↗

縫上一道針腳寬度約2mm的抽褶用縫線。→關於碎褶，請參考第100頁。

3

配合下襬的幅寬抽出碎褶，再用熨斗整燙。

4

將下襬的縫份用熨斗燙折起來。

5

塗抹布用接著劑。

6

將抽好碎褶的蕾絲暫時固定住。

7

縫合固定。

8

完成。

1. Prepare a lace that is about twice as long as the place where you want to gather the lace gathers.
2. Machine sew one line of gathering stitches 2mm in length in the upper. [refer to p.100 for gathering]
3. Gather the lace until the width fits the hem. Iron flat.   4. Fold the seam allowance of the hem with an iron.
5. Apply fabric glue.   6. Temporarily the lace on the hem.   7. Sew the hem.   8. Done.

m.
Frill
荷葉邊

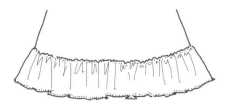

將荷葉邊上部的縫份折起後抽碎褶，再縫合到正面，這樣也很可愛哦！

1

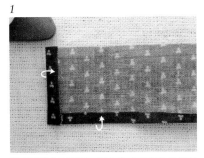

將荷葉邊的兩端及下襬的縫份用熨斗燙折起來。

2

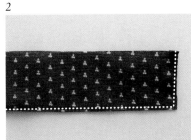

車縫壓線。

3

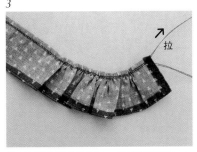

縫上兩道針腳寬度約2mm的抽褶用縫線。
→關於碎褶，請參考第100頁。

4

配合下襬的幅寬抽碎褶，把縫線綁好，再用熨斗燙壓固定。

5

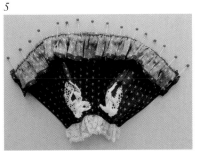

將荷葉邊正面相對合攏後，用珠針固定。

6

縫合起來。

7

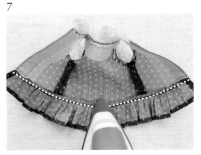

用熨斗將縫份倒向衣身那側。

8

車縫壓線。

9

完成。

1. Fold the seam allowance of the ruffled with an iron. *2.* Sew the edges.
3. Use a machine to sew two line of gathering stitch length 2mm on the upper seam allowance. [refer to p.100 for gathering]
4. Gather the fabric to match of the width of the hem. Iron flat. *5.* Pin the ruffle and hem. *6.* Sew the hem.
7. Iron the seam allowance toward to bodice. *8.* Sew reinforced stitches. *9.* Done.

n.
Drawn Work
抽花繡

將面料的橫線抽掉，製作出各種模樣的抽花繡。使用面料是平織薄棉布等較薄的平織布。

1

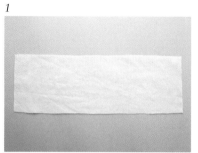

將想要抽花繡的面料，沿著布紋裁切成比紙型大上一圈的尺寸。

2

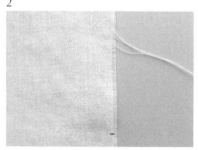

由布端抽掉幾根垂直的縱線。

3

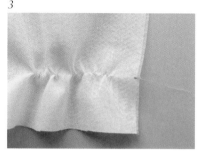

將想要製作抽花繡位置的橫線抽掉。

4

一次只抽掉一根線。

5

橫線抽掉幅寬約5mm左右之後的狀態。

6

將紙型複寫在面料上。

7

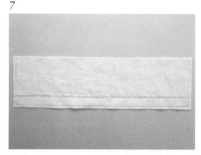

裁切成紙型大小，然後布端塗抹防綻液處理。

8

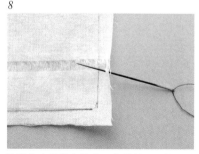

取單線穿過縫衣針後，將針刺入距離布端2mm左右的位置。

9

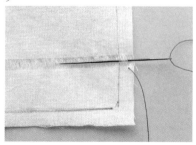

然後在針頭穿出面料位置的正上方，用針挑起約5mm幅寬的縱線。

1. Cut the dough you want to do drawn work along with the texture, slightly larger than the pattern.
2. Pull out several warps from the end.   3. Pull the weft thread where you want to apply the drawn work.
4. Pull the threads one by one.   5. Pull out the weft thread to about 5mm width.   6. Trace the pattern.   7. Cut and apply fray stopper liquid to all the edges.
8. Take one thread and pierce the needle about 2mm from the end.   9. Place your needle behind of the warp threads about 5mm width.

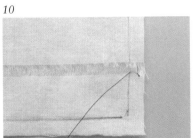

10

將縫線穿過後的狀態。

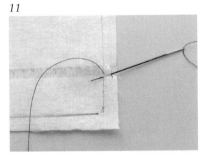

11

在與步驟9相同的位置再一次穿過縫針，然後在距離5mm的面料上穿過縫針。

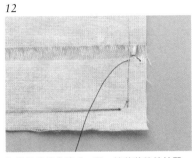

12

如此縫線就會繞成一圈，接著將縫線拉緊。

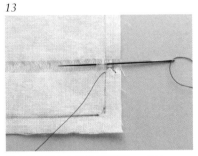

13

將出針位置上方的縱線挑起5mm左右。

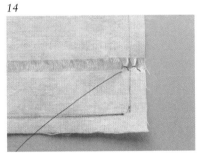

14

重覆9～12的步驟。

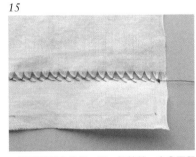

15

一直重覆到布端後，打一個線結。由背面觀看時是這樣的感覺。

16

由正面觀看時是這樣的感覺。

17

這是在圍裙的裙子部分加上抽花繡應用設計後的狀態。

18

罩衫的袖口距離較短，比較容易製作抽花繡，請試著挑戰看看吧！

10. Pull the needle through to the front. *11-12.* Put the needle through to the hole as step *9.* Pull out to the hem.
13-15. Repeat steps *9-12* to the full length. *16.* This is right side. *17.* The arrangement for apron dress hem.
18. For the sleeve, they are short and easy to challenge.

Gather
碎褶

這裡介紹的是如何將裙子、衣袖、袖口布拉攏成碎褶的製作方法。

1

調整縫紉機的刻度,將針腳的幅寬設定為2～2.5mm。

2

起縫點與止縫點不要使用回針縫,而是在縫份的正中央位置縫合。

3

兩端的縫線各保留15cm長,以便後續拉動抽褶。

4

在緊臨著第一道縫線的位置,平行縫上第二道縫線。

5

將上側縫線、下側縫線區分開來。

6

只拉動上側的2條縫線,抽出碎褶。抽褶的距離如果較長的話,由兩側一起拉動縫線;如果距離較短的話,先將單側縫線綁好後,再由另一側拉動縫線。

7

抽出所需長度的碎褶後,將上側縫線綁好,下側縫線也綁好。再將上下兩側縫線的另一端同樣綁好,就能將碎褶的幅寬固定下來。

8

將碎褶的間隔整理好之後,再用熨斗按壓固定。

9

抽碎褶的作業完成了。如果會很在意殘留在縫份上用來抽碎褶的縫線,將其拆掉也可以。

1. Set the machine to 2.5-3.0mm stitch.   *2.* Do not use back stitch as usual the start or end. Sew once along the edge.   *3.* Leave about 15cm of thread allowance on each side.   *4.* Sew a second line next to the first in the same way.   *5.* Separate both upper threads from the lower on each side.   *6.* Pull the upper threads while gathering the fabric.   *7.* When you have the desired width, knot all threads together on either side.   *8.* Iron the gathering to make it neat.   *9.* Cut away the thread allowance.

Snap
暗釦

安裝暗釦是衣服完成前的最後一道作業。本書的作品都是使用5mm的暗釦。

1

以雙線縫法，由暗釦的凹字形那側開始安裝。

2

每一個孔洞都要穿過2針。

3
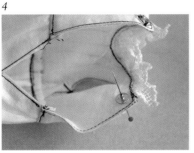

一邊確認後中心的重疊位置，一邊用珠針穿刺到凹字形暗釦的中心位置。

4

接下來將剛才那根珠針穿過到凸字形暗釦的中心位置。

5

保持珠針穿過的狀態，縫合固定兩個孔洞。

6

然後將珠針取下，把剩下來的孔洞都縫合固定。

7

下側的凹字形暗釦安裝完成後，將上側的凸字形暗釦扣上，一邊確認重疊的狀態，一邊將珠針如同步驟3一樣穿刺過去。

8

將凸字形暗釦也縫合固定後就完成了！

1. Start with socket side. Hand sew with two threads.　*2.* Insert two needles into each hole.
3-4. Insert a pin all the way through to mark opposite snap placement.
5. Sew through 2 holes with the pin inserted.　*6.* Then remove the pin and sew through holes.　*7-8.* Attach one more snaps.

---------------- 基本的A字連身裙 ----------------

M短版衣領f，衣袖j七分袖，
下襬m改為雙層

M短版衣領f，衣袖j七分袖，
下襬m改為雙層

M中版／衣身b，衣領e，
衣袖j七分袖，下襬m

M中版／衣身a，衣領g，
衣袖k，下襬l

M長版／衣領g，衣袖j五分袖

S長版／衣領e，衣袖j三分袖

S短版／衣身a，衣領g，下襬m，
衣袖k袖口布改為蕾絲

L中版／衣身b，衣領g，
衣袖j七分袖

L長版／衣身a，衣領f，
衣袖k，下襬m

---------------- 基本的罩衫 ----------------

M／衣領g，衣袖h，
下襬m

M／衣領e，衣袖k
下襬無碎褶

S／衣領f，衣袖k

S／衣身d，衣領e，
衣袖i，下襬m

L／衣領f，衣袖k

L／衣身a，衣領g，衣袖i，下襬m
袖口加上n

---------------- 基本的打褶連身裙 ----------------

M短版

M短版／衣領f，衣袖i
下襬加縫蕾絲

M中版／衣身c，
衣領g，衣袖j七分袖

L短版／衣身d，衣領e，衣袖i
下襬加縫蕾絲

L長版／衣領g，
衣袖j五分袖

S中版／衣身a，衣領g，衣袖h
下襬加縫蕾絲

S短版／衣身b，衣領f，
衣袖j五分袖

S長版／衣領g，衣袖k

S長版／衣身a，衣領f，
衣袖i，下襬n

"基本" 系列是由衣領、衣袖、下襬之中，選擇各尺寸的應用設計紙型來製作。

衣身的應用設計 *a.*蕾絲拼貼 *b.*蕾絲荷葉邊 *c.*蕾絲褶襯 *d.*蕾絲細褶*/*

衣領的應用設計 *e.*荷葉邊領 *f.*圓領 *g.*蕾絲領*/*

衣袖的應用設計 *h.*蕾絲袖 *i.*長袖 *j.*氣球袖 *k.*附袖口布衣袖*/*

下襬的應用設計 *l.*蕾絲碎褶 *m.*荷葉邊 *n.*抽花繡*/*

───── 百褶裙 ─────

M　　　　　　　　S　　　　　　　　S　　　　　　　　L

───── 連衫圍裙 ─────

S長版／下襬加縫蕾絲　　　　M長版　　　　M短版／衣身*d*，下襬*n*　　　　L短版　　　　L長版

───── 荷葉邊燈籠褲 ─────

M　　　　　M　　　　　S　　　　　L　　　　　L／下襬無荷葉邊

───── 夾克外套 ─────

M／蓬鬆衣袖　　　　S／長袖，衣領無荷葉邊　　　　L／蓬鬆衣袖，衣領無荷葉邊

───── 大衣 ─────

M／長袖，衣領荷葉邊　　　S／蓬鬆衣袖，衣領荷葉邊　　　L／長袖　　　L／長袖，西裝領

娃娃服飾縫紉書
HANON應用設計篇

作　　　者／藤井里美

翻　　　譯／楊哲群

發　 行　 人／陳偉祥

發　　　行／北星圖書事業股份有限公司

地　　　址／234新北市永和區中正路462號B1

電　　　話／886-2-29229000

傳　　　真／886-2-29229041

網　　　址／www.nsbooks.com.tw

E–MAIL／nsbook@nsbooks.com.tw

劃撥帳戶／北星文化事業有限公司

劃撥帳號／50042987

製版印刷／皇甫彩藝印刷股份有限公司

出 版 日／2021年01月

I S B N／978-957-9559-63-8

定　　　價／400

國家圖書館出版品預行編目(CIP)資料

HANON：娃娃服飾縫紉書. 應用設計篇／藤井里美作；

楊哲群翻譯. -- 新北市：北星圖書，2021.01

　　面；　公分

ISBN 978-957-9559-63-8（平裝）

1.洋娃娃 2.手工藝

426.78　　　　　　　　　　　109014332

臉書粉絲專頁　　　　LINE 官方帳號

DOLL SEWING BOOK

HANON

───── 應用設計篇 ─────

Pattern

這裡刊載的紙型基本上都是100%原尺寸的大小。

S尺寸的紙型是紅色

M尺寸的紙型是黑色

L尺寸的紙型是藍色

各尺寸都有區分不同顏色，請將想要製作的尺寸的紙型以原尺寸複印後，裁剪下來使用。

───── 紙型的複製方法 ─────

將紙型放置於面料的背面，使用粉土筆或是布用複寫紙，先以粗線
條沿著紙型描繪「完成線」，再於線條外側描繪細線條的「縫份線」。
「縫份線」是面料剪裁的位置。
粗線條的「完成線」則是縫合時的縫線位置。

←→　　　　這個箭頭指的是布紋的「縱向」方向(面料上有布邊的那側是縱向)。

▶　　　　　這個三角記號是「開口止點」記號。請一定要複寫這個記號。

──　　　　這是指定「蕾絲縫製止點」以及碎褶位置的記號。請一定要複寫這個記號。

○∽○　　　這是抽碎褶範圍的記號。

「左右×各1」　將紙型直接放在面料上1張，將紙型翻至背面左右翻轉1張，合計製作2張。

「表裡×各1」　將紙型直接放在面料正面1張，直接放在面料背面1張，合計製作2張。

「×2」　　　　將紙型直接放在面料上製作2張。

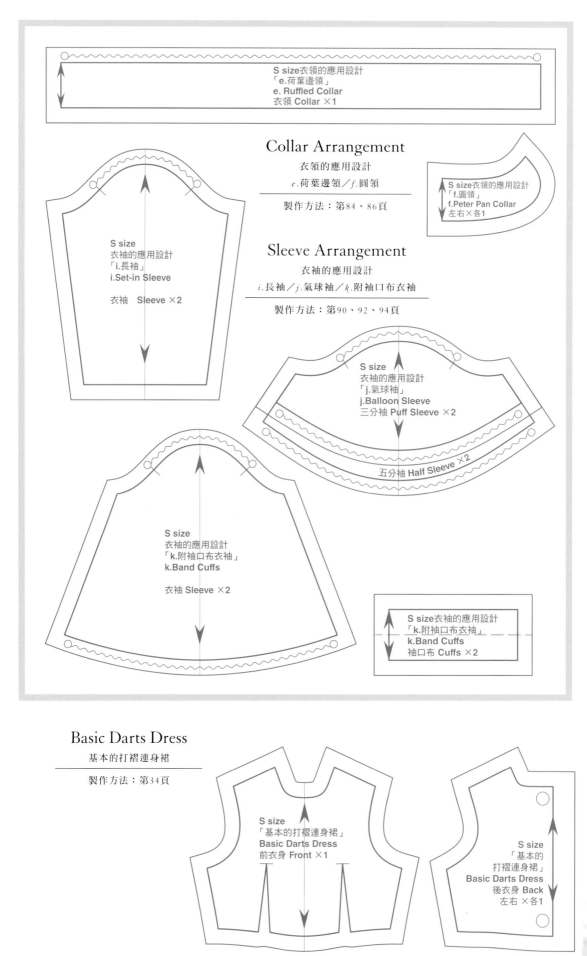

S size衣領的應用設計
「e.荷葉邊領」
e. Ruffled Collar
衣領 Collar ×1

Collar Arrangement
衣領的應用設計

*e.*荷葉邊領／*f.*圓領

製作方法：第84、86頁

S size衣領的應用設計
「f.圓領」
f.Peter Pan Collar
左右×各1

S size
衣袖的應用設計
「i.長袖」
i.Set-in Sleeve

衣袖 Sleeve ×2

Sleeve Arrangement
衣袖的應用設計

*i.*長袖／*j.*氣球袖／*k.*附袖口布衣袖

製作方法：第90、92、94頁

S size
衣袖的應用設計
「j.氣球袖」
j.Balloon Sleeve
三分袖 Puff Sleeve ×2

五分袖 Half Sleeve ×2

S size
衣袖的應用設計
「k.附袖口布衣袖」
k.Band Cuffs

衣袖 Sleeve ×2

S size衣袖的應用設計
「k.附袖口布衣袖」
k.Band Cuffs
袖口布 Cuffs ×2

Basic Darts Dress
基本的打褶連身裙

製作方法：第34頁

S size
「基本的打褶連身裙」
Basic Darts Dress
前衣身 Front ×1

S size
「基本的
打褶連身裙」
Basic Darts Dress
後衣身 Back
左右 ×各1

S size
「百褶裙」
Tiered Skirt
裙片上 Skirt ×1

S size
「百褶裙」
Tiered Skirt
裙片下 Frill ×1

摺雙

Tiered Skirt
百褶裙

製作方法：第42頁

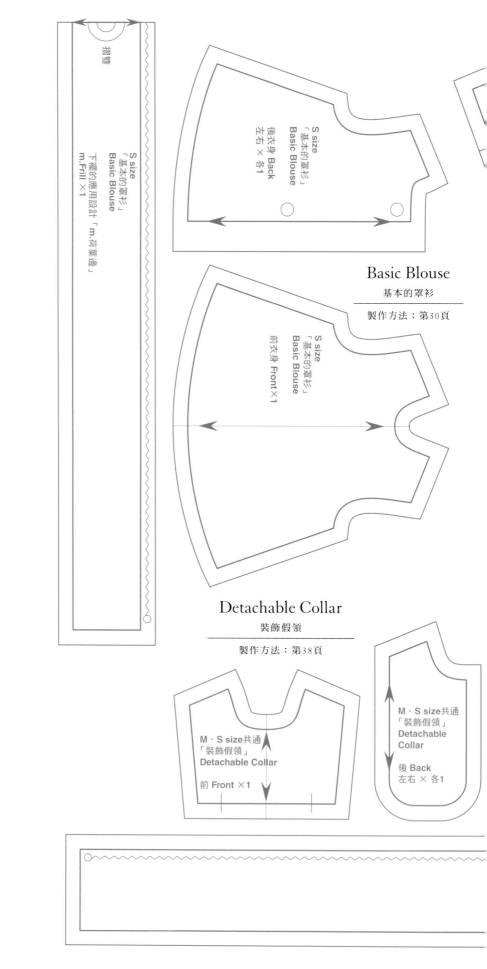

摺雙

S size
「基本的罩衫」
Basic Blouse

下襬的應用設計「m.荷葉邊」
m.Frill ×1

S size
「基本的罩衫」
Basic Blouse
後衣身 Back
左右 ×各1

Basic Blouse
基本的罩衫

製作方法：第30頁

S size
「基本的罩衫」
Basic Blouse
前衣身 Front×1

Detachable Collar
裝飾假領

製作方法：第38頁

M、S size共通
「裝飾假領」
Detachable Collar

前 Front ×1

M、S size共通
「裝飾假領」
Detachable
Collar

後 Back
左右 × 各1

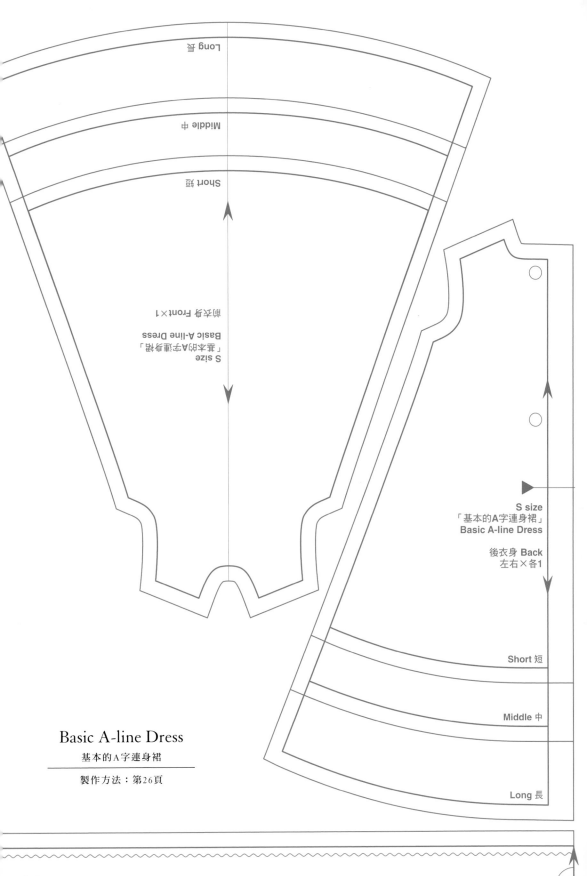

Long 套

Middle 中

Short 短

S size
「基本的A字連身裙」
Basic A-line Dress

前衣身 Front×1

S size
「基本的A字連身裙」
Basic A-line Dress

後衣身 Back
左右×各1

Short 短

Middle 中

Long 長

Basic A-line Dress

基本的A字連身裙

製作方法：第26頁

S size
「基本的A字連身裙」
Basic A-line Dress

下襬的應用設計「m.荷葉邊」
m.Frill ×1

摺雙

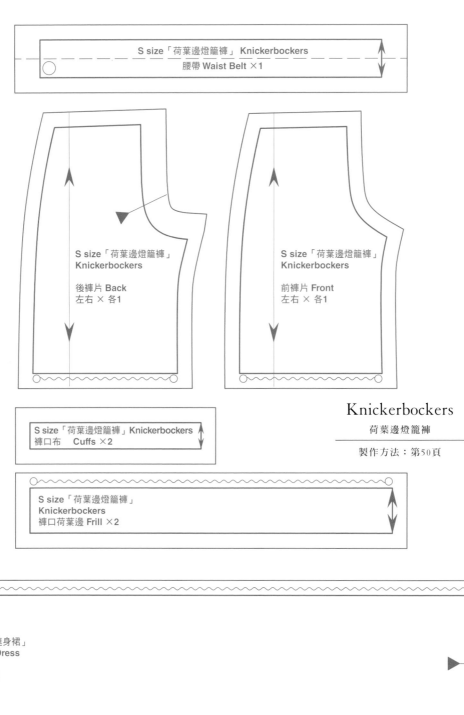

S size「荷葉邊燈籠褲」 Knickerbockers
腰帶 Waist Belt ×1

S size「荷葉邊燈籠褲」
Knickerbockers

後褲片 Back
左右 × 各1

S size「荷葉邊燈籠褲」
Knickerbockers

前褲片 Front
左右 × 各1

Knickerbockers
荷葉邊燈籠褲

製作方法：第50頁

S size「荷葉邊燈籠褲」 Knickerbockers
褲口布　Cuffs ×2

S size「荷葉邊燈籠褲」
Knickerbockers
褲口荷葉邊 Frill ×2

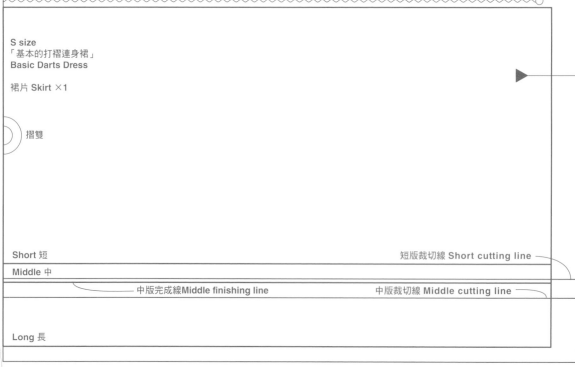

S size
「基本的打褶連身裙」
Basic Darts Dress

裙片 Skirt ×1

摺雙

Short 短

短版裁切線 Short cutting line

Middle 中

中版完成線Middle finishing line　　　　中版裁切線 Middle cutting line

Long 長

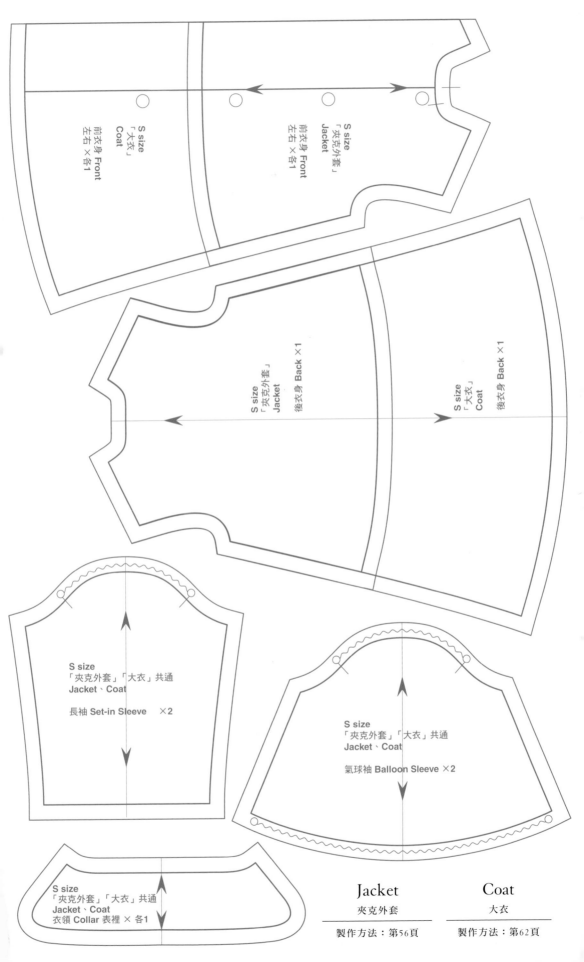

S size
「大衣」Coat
前衣身 Front
左右 ×各1

S size
「夾克外套」Jacket
前衣身 Front
左右 ×各1

S size
「夾克外套」Jacket
後衣身 Back ×1

S size
「大衣」Coat
後衣身 Back ×1

S size
「夾克外套」「大衣」共通
Jacket、Coat

長袖 Set-in Sleeve ×2

S size
「夾克外套」「大衣」共通
Jacket、Coat

氣球袖 Balloon Sleeve ×2

S size
「夾克外套」「大衣」共通
Jacket、Coat
衣領 Collar 表裡 ×各1

Jacket
夾克外套

製作方法：第56頁

Coat
大衣

製作方法：第62頁

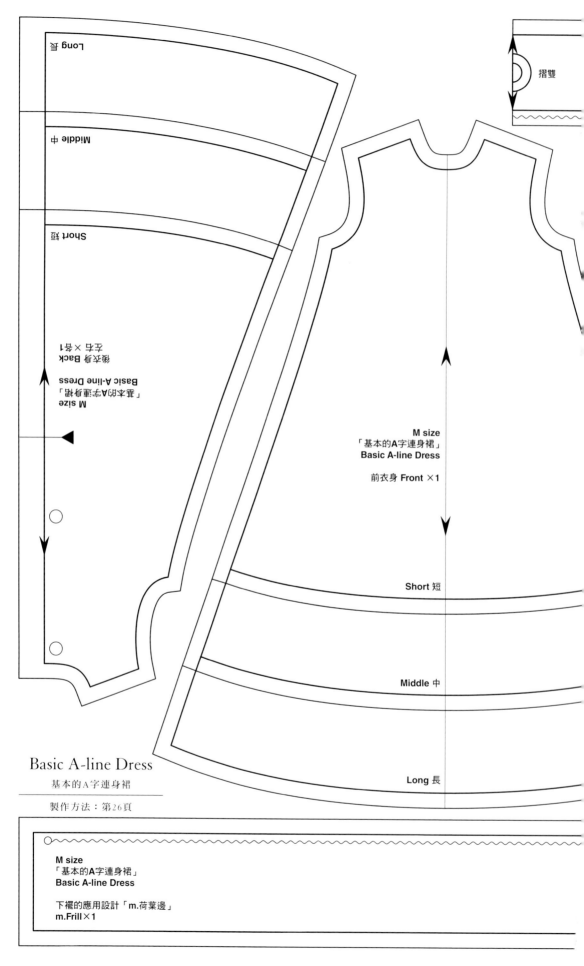

Long 長

Middle 中

Short 短

摺雙

後衣身 Back
布紋 × 各1
「基本的A字連身裙」
Basic A-line Dress
M size

M size
「基本的A字連身裙」
Basic A-line Dress

前衣身 Front ×1

Short 短

Middle 中

Long 長

Basic A-line Dress
基本的A字連身裙

製作方法：第26頁

M size
「基本的A字連身裙」
Basic A-line Dress

下襬的應用設計「m.荷葉邊」
m.Frill ×1

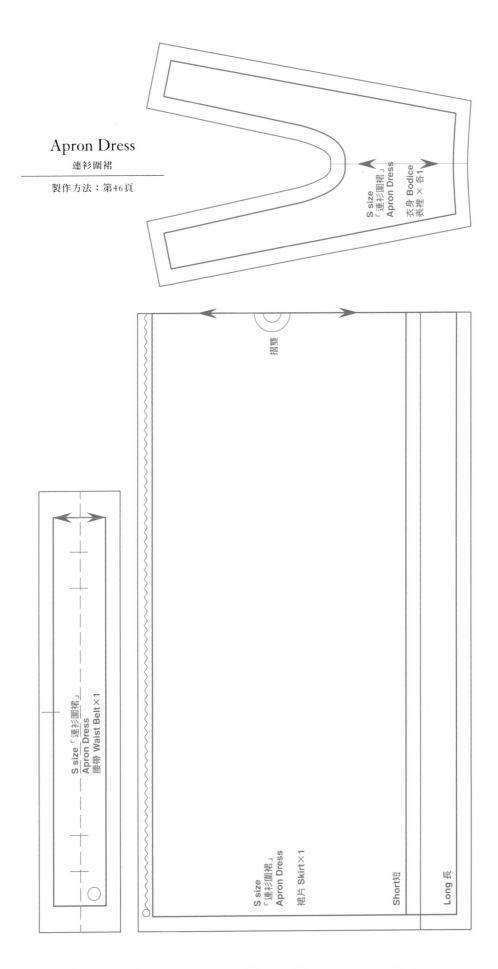

Apron Dress

連衫圍裙

製作方法：第46頁

S size「連衫圍裙」
Apron Dress
衣身 Bodice
表裡 × 各1

摺雙

S size「連衫圍裙」
Apron Dress
腰帶 Waist Belt×1

S size
「連衫圍裙」
Apron Dress
裙片 Skirt×1

Short 短

Long 長

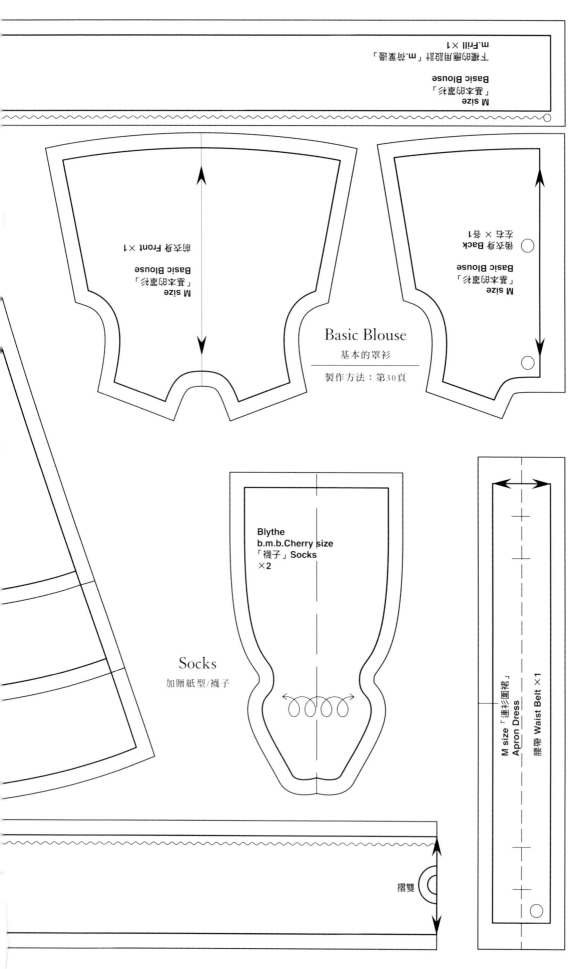

m.Frill ×1
下襬的蕾絲用厚紙樣「m.荷葉邊」
Basic Blouse
M size
「基本的罩衫」

後衣身 Back ×1
Basic Blouse
M size
「基本的罩衫」

前衣身 Front ×1
Basic Blouse
M size
「基本的罩衫」

Basic Blouse
基本的罩衫

製作方法：第30頁

Blythe
b.m.b.Cherry size
「襪子」Socks
×2

Socks
加贈紙型/襪子

M size 「連衫圍裙」
Apron Dress

腰帶 Waist Belt ×1

摺雙

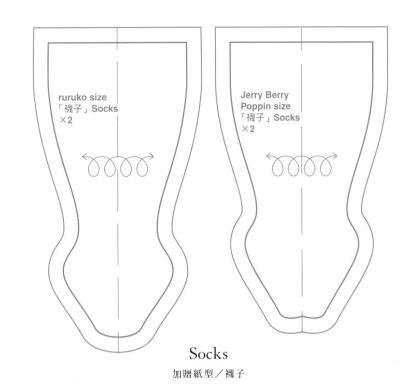

Socks

加贈紙型／襪子

Hat

帽子

製作方法：第68頁

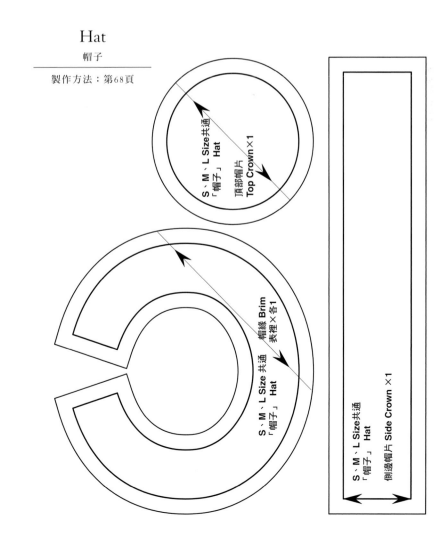

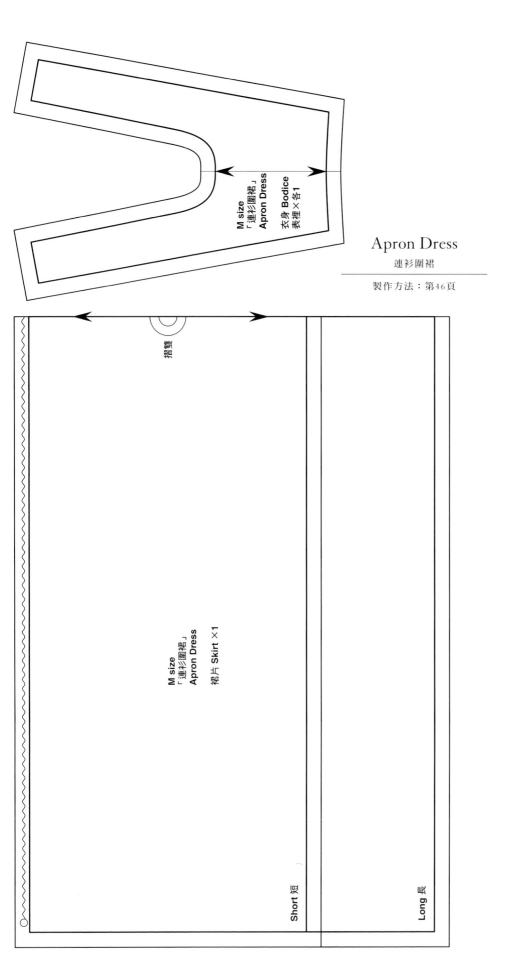

Apron Dress
連衫圍裙

製作方法：第46頁

M size
「連衫圍裙」
Apron Dress
衣身 **Bodice**
表裡×各1

M size
「連衫圍裙」
Apron Dress
裙片 **Skirt** ×1

摺雙

Short 短

Long 長

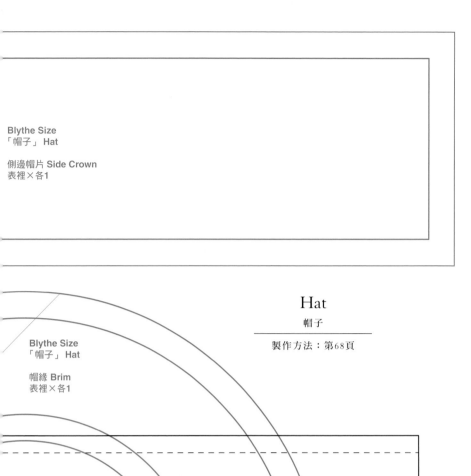

Blythe Size
「帽子」 Hat

側邊帽片 Side Crown
表裡×各1

Hat

帽子

製作方法：第68頁

Blythe Size
「帽子」 Hat

帽緣 Brim
表裡×各1

摺雙

Knickerbockers

荷葉邊燈籠褲

製作方法：第50頁

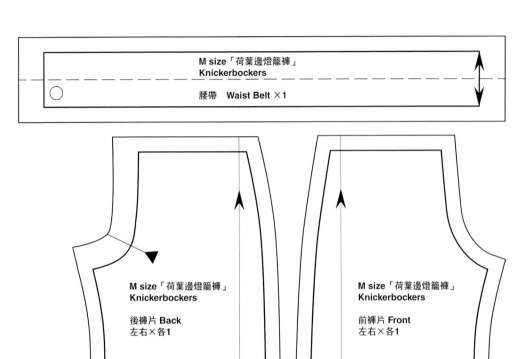

M size「荷葉邊燈籠褲」
Knickerbockers

腰帶　Waist Belt ×1

M size「荷葉邊燈籠褲」
Knickerbockers

後褲片 Back
左右×各1

M size「荷葉邊燈籠褲」
Knickerbockers

前褲片 Front
左右×各1

M size「荷葉邊燈籠褲」
Knickerbockers
褲口布 Cuffs　×2

M size「荷葉邊燈籠褲」
Knickerbockers
褲口荷葉邊 Frill　×2

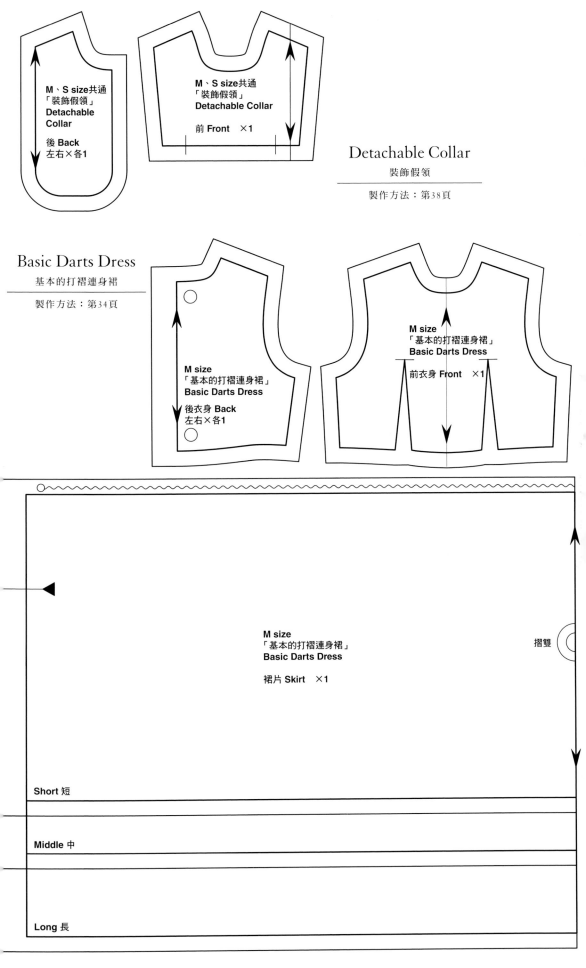

M、S size共通
「裝飾假領」
Detachable Collar

後 Back
左右×各1

M、S size共通
「裝飾假領」
Detachable Collar

前 Front ×1

Detachable Collar
裝飾假領

製作方法：第38頁

Basic Darts Dress
基本的打褶連身裙

製作方法：第34頁

M size
「基本的打褶連身裙」
Basic Darts Dress

後衣身 Back
左右×各1

M size
「基本的打褶連身裙」
Basic Darts Dress

前衣身 Front ×1

M size
「基本的打褶連身裙」
Basic Darts Dress

裙片 Skirt ×1

摺雙

Short 短

Middle 中

Long 長

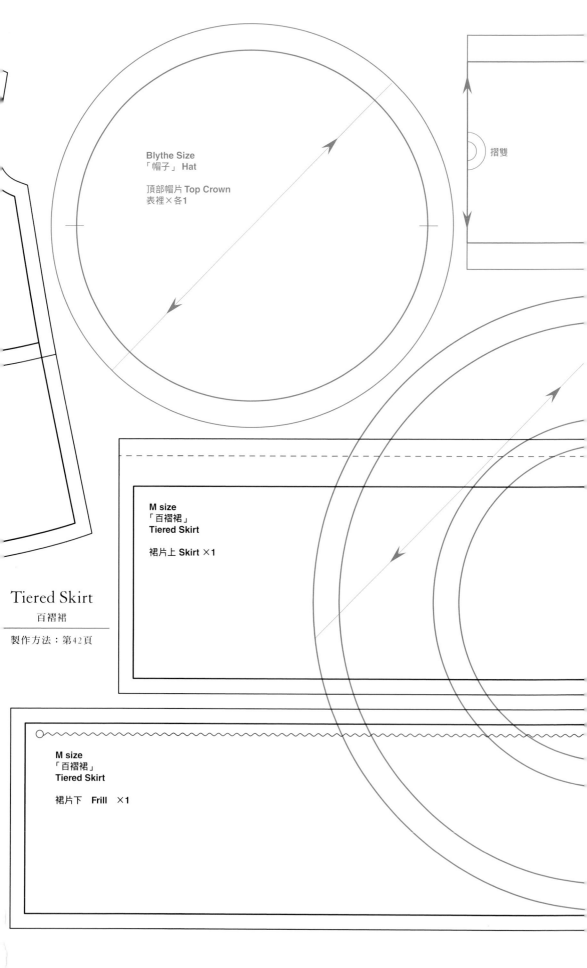

Blythe Size
「帽子」 Hat

頂部帽片 **Top Crown**
表裡×各1

摺雙

M size
「百褶裙」
Tiered Skirt

裙片上 **Skirt** ×1

Tiered Skirt
百褶裙

製作方法：第42頁

M size
「百褶裙」
Tiered Skirt

裙片下　**Frill**　×1

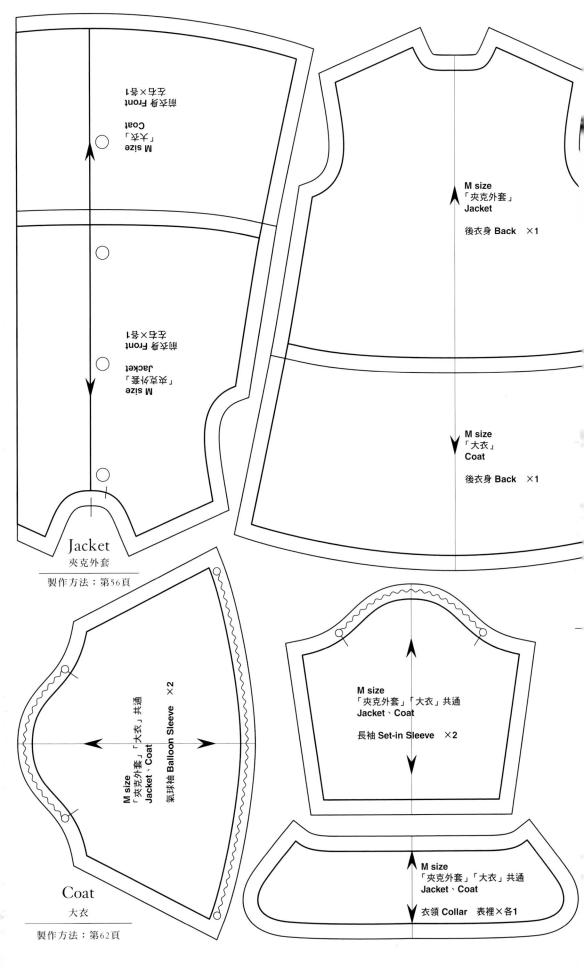

後衣身 Front
名布×各1
Coat
「大衣」
M size

後衣身 Front
名布×各1
Jacket
「夾克外套」
M size

M size
「夾克外套」
Jacket

後衣身 Back ×1

M size
「大衣」
Coat

後衣身 Back ×1

Jacket
夾克外套

製作方法：第56頁

M size
「夾克外套」「大衣」共通
Jacket、Coat

長袖 Set-in Sleeve ×2

M size
「夾克外套」「大衣」共通
Jacket、Coat
氣球袖 Balloon Sleeve ×2

Coat
大衣

製作方法：第62頁

M size
「夾克外套」「大衣」共通
Jacket、Coat

衣領 Collar 表裡×各1

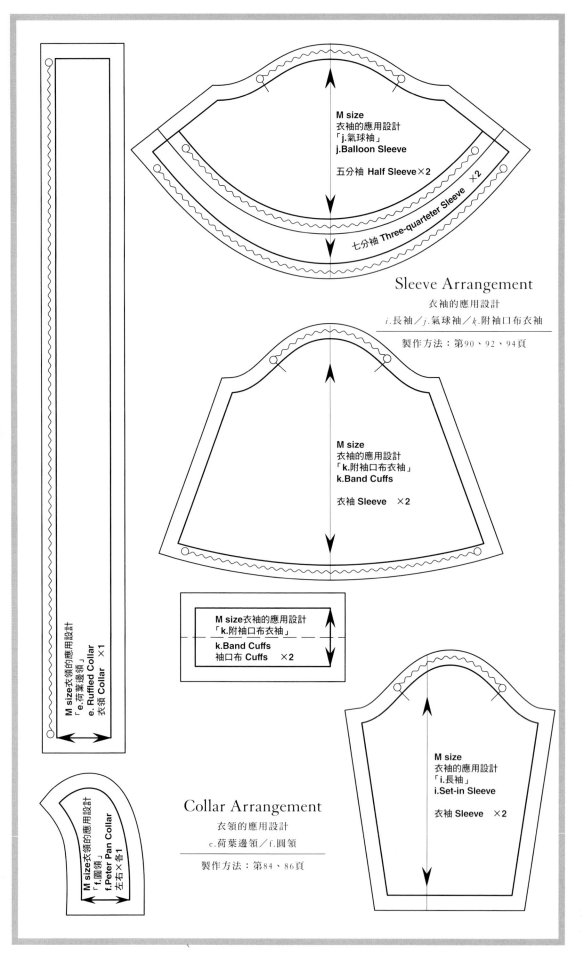

M size
衣袖的應用設計
「**j.**氣球袖」
j.Balloon Sleeve

五分袖 Half Sleeve ×2

七分袖 Three-quarteter Sleeve ×2

Sleeve Arrangement
衣袖的應用設計
*i.*長袖／*j.*氣球袖／*k.*附袖口布衣袖

製作方法：第90、92、94頁

M size
衣袖的應用設計
「**k.**附袖口布衣袖」
k.Band Cuffs

衣袖 Sleeve ×2

M size衣袖的應用設計
「**k.**附袖口布衣袖」
k.Band Cuffs
袖口布 Cuffs ×2

M size衣領的應用設計
「**e.**荷葉邊領」
e. Ruffled Collar
衣領 Collar ×1

M size衣領的應用設計
「**f.**圓領」
f.Peter Pan Collar
左右×各1

Collar Arrangement
衣領的應用設計
e.荷葉邊領／f.圓領

製作方法：第84、86頁

M size
衣袖的應用設計
「**i.**長袖」
i.Set-in Sleeve

衣袖 Sleeve ×2